Education and Ethics in the Life Sciences

EDITED BY BRIAN RAPPERT

Education and Ethics in the Life Sciences

Strengthening the Prohibition of Biological Weapons

EDITED BY BRIAN RAPPERT

THE AUSTRALIAN NATIONAL UNIVERSITY
E PRESS

Centre for Applied Philosophy and Public Ethics
An Australian Research Council Special Research Centre

Practical Ethics and Public Policy Monograph 1

Series Editor: Michael J. Selgelid

ANU E PRESS

Published by ANU E Press
The Australian National University
Canberra ACT 0200, Australia
Email: anuepress@anu.edu.au
This title is also available online at: http://epress.anu.edu.au/education_ethics_citation.html

National Library of Australia
Cataloguing-in-Publication entry

Title: Education and ethics in the life sciences : strengthening the prohibition of biological weapons / edited by Brian Rappert.

ISBN: 9781921666384 (pbk.) 9781921666391 (ebook)

Series: Practical ethics and public policy ; no. 1.

Subjects: Biological arms control--Verification.
Biosecurity.
Bioethics.
Communicable diseases--Prevention.

Other Authors/Contributors: Rappert, Brian.

Dewey Number: 327.1745

Cover design and layout by ANU E Press

Contents

PART 3: THE WAYS FORWARD

Acknowledgments

Thanks to all those who contributed to this book through their chapters and suggestions. Grants from the Alfred P. Sloan Foundation and the Wellcome Trust enabled the editor to undertake the work of preparing this volume.

Contributors

Michael Barr is Research Councils UK Fellow and Lecturer at Newcastle University, England. His research focuses on the implications of the rise of China, especially Sino-Western relations. He was Visiting Fellow at the Chinese Academy of Medical Sciences in 2008 and has lectured widely in China on bioethics and biosecurity issues.

Louise Bezuidenhout completed her PhD in cardio-thoracic surgery at the University of Cape Town, South Africa, and has worked as a post-doctoral scientist in the department of cardiovascular sciences at the University of Edinburgh, UK. In 2008 she was the recipient of an EU scholarship for the Erasmus Mundus Masters of Bioethics programme run by the Catholic University of Leuven, Radboud University Nijmegen and the University of Padova. These dual interests in science and bioethics led her to enrol for a PhD in sociology at the University of Exeter under Professor Brian Rappert (to be completed in 2012). Louise is the recipient of a scholarship from the Wellcome Trust to work on the collaborative project on building sustainable capacity in dual-use bioethics. Her research focus is on the conceptualisation of the dual-use debate by life scientists in developing countries, and how different social surroundings affect involvement in dual-use debates.

Nancy D. Connell is Professor of Medicine at the University of Medicine and Dentistry of New Jersey (UMDNJ)-New Jersey Medical School. She is also director of the UMDNJ Center for BioDefense, which was established in 1999 and is the recipient of $11.5 million in congressional recommendations (2000–06) for research into the detection and diagnosis of biological warfare agents and biodefence preparedness. Dr. Connell is also director of the Biosafety Level 3 Facility of UMDNJ's Center for the Study of Emerging and Re-emerging Pathogens and chairs the Recombinant DNA Subcommittee of the university's Institutional Biosafety Committee. She has worked with several international programmes on dual-use issues. She is past chair of the National Institutes of Health's Center for Scientific Review Study Section HIBP (Host Interactions with Bacterial Pathogens), which reviews bacterial-pathogenesis submissions to the National Institute of Allergy and Infectious Diseases. She is current chair of the F13 infectious diseases and microbiology fellowship panel. Dr. Connell's involvement in biological-weapons control began in 1984, when she was chair of the Committee on the Military Use of Biological Research, a subcommittee of the Council for Responsible Genetics, based in Cambridge, Massachusetts. Dr.

Connell received her PhD in microbial genetics from Harvard University. Her major research focus is the interaction between Mycobacterium tuberculosis and the macrophage.

Malcolm Dando is Professor of International Security at the University of Bradford. A biologist by original training, his main research interest is in the preservation of the prohibitions embodied in the Chemical Weapons Convention and the Biological Weapons Convention at a time of rapid scientific and technological change in the life sciences. His recent publications include *Deadly Cultures: Biological Weapons Since 1945* (Harvard University Press, 2006), which he edited with Mark Wheelis and Lajos Rozsa.

Christian Enemark is Senior Lecturer in the Centre for International Security Studies, University of Sydney, and founding Director (Sydney) of the National Centre for Biosecurity. His research and teaching interests include the security implications of infectious-disease threats, including biological weapons, and the ethics and laws of armed conflict. Prior to completing a PhD at the Strategic and Defence Studies Centre, Australian National University (ANU), Christian worked as a policy advisor in the Parliament and Attorney General's Department of New South Wales. In 2007 and 2008 he was a Visiting Fellow of the ANU John Curtin School of Medical Research and a member of the National Consultative Committee on International Security Issues (Department of Foreign Affairs and Trade). Christian is presently a Visiting Fellow at the Centre for Applied Philosophy and Public Ethics at ANU, and Chief Investigator for a three-year Australian Research Council project on 'Infectious Disease, Security and Ethics'. He was Principal Investigator (Australia), funded by the Alfred P. Sloan Foundation, for a 2009 pilot programme of seminars on the dual-use dilemma in the life sciences.

David Friedman holds a MSc degree in microbiology from Tel Aviv University (1972) and a PhD in Immunology from the Weizmann Institute, Rehovot (1976). After one year of post-doctoral training (1977) in the Max-Planck Institute for Immunobiology, Freiburg, Germany, he joined the Israeli Defense Forces (IDF). For nearly 25 years, Dr. Friedman served in the IDF and then in the Israeli Ministry of Defense (IMOD) and retired as a colonel (1992). Dr. Friedman served mainly in the R&D directorate, and was responsible for planning and R&D projects in the field of chemical/biological defense. In the IDF he was Head of the Chem/Bio Protection Division and in the IMOD he served as Special Assistant for Bio/Chem defense. In 2004, Dr. Friedman joined the Jaffee Center for Strategic Studies (JCSS), Tel-Aviv University, now called Institute for National Security Studies (INSS),. Dr. Friedman is currently a senior research associate in INSS and his research focuses on WMD, non-conventional terrorism, with particular attention to strategies for non-proliferation, counter-proliferation of chemical and biological weapons, focusing mainly on combating bioterrorism. In addition

Dr. Friedman is an adviser to organisations such as the Israeli Academy of Sciences, the National Security Council, the Ministry of Defense, the Ministry of Foreign Affairs and the Ministry of Health.

François Garraux works as a policy adviser on arms control and disarmament in the Swiss Federal Department of Defence, Civil Protection and Sport. Prior to this, he worked in the communication and media branch of the same Department, as well as abroad for the Swiss Federal Department of Foreign Affairs. François Garraux is a staff officer in the Swiss Armed Forces. He graduated in contemporary Asian studies at the University of Amsterdam, as well as in the history and science of media at the University of Bern.

Jane Johnson is a post-doctoral research fellow in clinical and public-health ethics at the Philosophy Department of Macquarie University, where she pursues a wide range of philosophical and research interests. Her PhD (completed at the University of Sydney) examined, using the philosophical resources of Kant and Hegel, the problem of justifying legal punishment and she is currently working on a project regarding the ethical issues generated by innovative surgery. Strong interests in nonhuman animals and teaching have led Jane into research on the ethical and epistemological implications of using animals in biomedicine, and philosophical pedagogy.

Giulio M. Mancini is a researcher and programme officer at the Landau Network-Centro Volta (LNCV), Italy, on nuclear and biological-weapons proliferation, prevention and disarmament, and biosafety/biosecurity enhancement tools and policies. He coordinates the LNCV project on Biosecurity and Dual Use Education. He holds an MSc in International Relations and European Integration and a BSc in Political Science, both from the Catholic University of Milan.

Brendan McCluskey was appointed the Executive Director of the Office of Emergency Management and Occupational Health and Safety at the University of Medicine and Dentistry of New Jersey (UMDNJ) in 2006, and directs security for the university's Biosafety Level 3 laboratories. He had previously held positions as Deputy Director of the Center for BioDefense (2001–04) and as Acting Director of the Chemical, Biological, Radiological, Nuclear, and Explosive Center for Training and Research (2004–06) at the university. He has served as a member of the Governor's Task Force on Campus Safety (New Jersey) since 2007. Mr. McCluskey is a Certified Emergency Manager and serves as Chair of the Universities and Colleges Caucus of the International Association of Emergency Managers. He was appointed an Assistant Professor in the Graduate School of Biomedical Sciences at UMDNJ in 2002, where he teaches courses on bioterrorism, weapons of mass destruction, and homeland security. Until 2009, Mr. McCluskey was also an Assistant Professor at Kean University, where he taught courses in public administration, bioterrorism, and public-health policy. Mr. McCluskey received his J.D. (2006) from Rutgers University School of Law, and his BA (1997) and MPA. (2001) from Kean University.

Masamichi Minehata is a post-doctoral research fellow at the Bradford Disarmament Research Centre (BDRC) of the University of Bradford's Department of Peace Studies. Under the auspices of the UK Prime Minister's Initiative on International Research Cooperation (British Council), from 2008 to the beginning of 2010 he worked on the joint project between the BDRC and the National Defense Medical College of Japan to develop an online educational module resource for life scientists about dual-use issues. His current research focuses on the education of life scientists about these issues in Asian countries under the Alfred P. Sloan Foundation of the United States.

Brian Rappert is an Associate Professor of Science, Technology and Public Affairs in the Department of Sociology and Philosophy at the University of Exeter, UK. His long-term interest has been the examination of how choices can be, and are, made about the adoption and regulation of security-related technologies, particularly in conditions of uncertainty and disagreement. His recent books include *Controlling the Weapons of War* (Routledge, 2006), *Biotechnology, Security and the Search for Limits* (Palgrave, 2007), *Technology & Security* (ed., Palgrave, 2007), *Biosecurity* (co-ed., Palgrave, 2009). His most recent, *Experimental Secrets* (UPA, 2009), offers a novel approach for investigating and writing about the place of absences in social inquiry.

James Revill is a final year doctoral candidate studying the Evolution of the Biological and Toxins Weapons Convention at Bradford's Disarmament Research Centre. Over the course of completing his thesis he has been engaged in a number of research projects focused on biological weapons, biosecurity and development, but also issues related to South Asian security through his role in the Pakistan Security Research Unit at Bradford. However, he has been particularly active in the sphere of biosecurity education, an interest that has been pursued through employment as a Research Fellow by the Landau Network Centro Volta in Italy and as a Visiting Research Fellow with Bradford Disarmament Research Centre in the UK.

Michael J. Selgelid earned a BS in Biomedical Engineering from Duke University and a PhD in Philosophy from the University of California, San Diego. He is a Senior Research Fellow in the Centre for Applied Philosophy and Public Ethics (CAPPE) at the Australian National University, where he is also Director of a World Health Organisation (WHO) Collaborating Centre for Bioethics and Deputy Director of the National Centre for Biosecurity. He has held previous appointments at the University of Sydney and the University of the Witwatersrand in Johannesburg, South Africa. His research focuses on ethical issues associated with infectious disease and genetics. He co-authored *Ethical and Philosophical Consideration of the Dual-Use Dilemma in the Biological Sciences* (Springer, 2008) and co-edited *Ethics and Infectious Disease* (Blackwell, 2006).

Nariyoshi Shinomiya is Professor of the Department of Integrative Physiology and Bio-Nano Medicine at the National Defense Medical College (NDMC), Japan. After he finished the Graduate Course of Medicine of NDMC in 1991, he worked as Associate Professor at the Department of Biology, NDMC (1993–97) and at the Department of Microbiology, NDMC (1997–2007), and then moved to the present position. His specialty includes microbiology and immunology, molecular oncology, diving medicine, and bioethics. His recent work on bioethics has focused on the dual-use issues and education for life scientists.

Judi Sture is the Head of the Graduate School at the University of Bradford, England, where she leads two doctoral research training programmes. She lectures in Research Ethics and Research Methodology and is closely involved in devising and developing postgraduate and ethics policy and practice at the University and beyond. As a member of the Wellcome Trust Dual-Use Bioethics Group and associate member of the Bradford Disarmament Research Centre she is engaged with colleagues from a number of UK and overseas universities in developing a bioethics approach to counter biosecurity threats in the life sciences. Judi holds a BSc(Hons) in Archaeology (University of Bradford), in which she specialised in the study of human skeletal remains, and a PhD (University of Durham) in biological anthropology, focusing on environmental associations with human birth defects. Her research in biological anthropology continues, including further work on developmental defects. She is currently engaged in analysis of skeletal remains held at the Museum of London and is working with the British Association for Biological Anthropology and Osteoarchaeology on developing ethical practice in the profession.

Simon Whitby is a Research Councils UK-sponsored Senior Research Fellow and the Director of Bradford Disarmament Research Centre at the University of Bradford. His research interests include the examination of the characteristics and use of military programmes that target crops with biochemical agents. His 2001 book *Biological Warfare Against Crops* represents the first substantive study of state-run activities designed to target food crops with biological warfare agents.

Joy Yueyue Zhang is a Wellcome Trust-funded PhD candidate at the BIOS Centre of the London School of Economics and Political Science. Her thesis investigates Chinese regulation of stem-cell research in the context of cosmopolitanisation. She completed her first degree in Medicine at Peking University.

List of Abbreviations

AAAS	American Association for the Advancement of Science
AAMC	American Association of Medical Colleges
AAU	Association of American Universities
ANU	Australian National University
BBSRC	Biotechnology and Bioscience Research Council
BDRC	Bradford Disarmament Research Centre
BSATs	Biological Select Agents and Toxins
BSL	biosafety level
BTWC	Biological and Toxin Weapons Convention
BW	biological warfare (or weapons)
CAS	Chinese Academy of Sciences
CASS	Chinese Academy of Social Sciences
CBRN	chemical, biological, radiological and nuclear
CDC	Centers for Disease Control
COGR	Council on Government Relations
CSR	corporate social responsibility
CTR	Cooperative Threat Reduction
CW	chemical weapons
CWC	Chemical Weapons Convention
DDPS	Federal Department of Defence, Civil Protection and Sport (Switzerland)
DHS	Department of Homeland Security (US)
DHHS	Department of Health and Human Services (US)
DoD	Department of Defense (US)

EMR	education module resource
EU	European Union
FASEB	Federation of American Societies for Experimental Biology
FBI	Federal Bureau of Investigation
FDFA	Federal Department of Foreign Affairs (Switzerland)
FOPH	Federal Office of Public Health (Switzerland)
HEFCE	Higher Education Funding Council
IAMP	InterAcademy Medical Panel
IAP	InterAcademy Panel
ICRC	International Committee of the Red Cross
ISP	intersessional process
IUPAC	International Union of Pure and Applied Chemistry
IUMS	International Union of Microbiological Societies
JBA	Japan Bioindustry Association
LNCV	Landau Network Centro Volta (Italy)
METI	Ministry of Economy, Trade and Industry (Japan)
MEXT	Ministry of Education, Culture, Sports, Science and Technology (Japan)
MRC	Medical Research Council (UK)
MOST	Ministry of Science and Technology (China)
NAS	National Academy of Sciences (US)
NASILGC	Association of Public and Land-grant Universities (US)
NBC	nuclear, biological, and chemical
NCB	National Centre for Biosecurity (Switzerland)
NDMC	National Defense Medical College (Japan)
NGO	non-governmental organisation

NIH	National Institutes of Health (US)
NIID	National Institute of Infectious Diseases (Japan)
NRC	National Research Council
NSABB	National Science Advisory Board for Biosecurity
OECD	Organisation for Economic Co-operation and Development
ORI	Office of Research Integrity
OSHA	Occupational Safety and Health Association
PHS	public-health system
QAA	Quality Assurance Agency
R&D	research and development
RCR	Responsible Conduct of Research
RCUK	Research Councils UK
RISTEX	Research Institute of Science and Technology for Society (Japan)
S&T	science and technology
SCJ	Science Council of Japan
SECO	State Secretariat for Economic Affairs (Switzerland)
SER	State Secretariat for Education and Research (Switzerland)
SSBAs	Security-Sensitive Biological Agents (Australia)
UK	United Kingdom
UN	United Nations
US	United States of America
WHO	World Health Organisation
WMD	weapons of mass destruction

PART 1

ETHICS, EDUCATION AND THE LIFE SCIENCES

Introduction: Education as...

BRIAN RAPPERT

The history of modern science and technology is a story that cannot be told without attending to the military, destructive, and violent purposes motivating the search for new knowledge and devices.[1] At certain times, the pace or novelty of developments have been seen as demanding social debate. The construction of atomic and nuclear weapons is perhaps the exemplary case where pause and concern has been evident about what the capabilities of some mean for the many.

At the start of the twenty-first century, warnings have been raised in some quarters about how — by intent or by mishap — advances in biotechnology and related fields could aid the spread of disease. Science academies, medical organisations, government commissions and security analysts, as well as individual researchers, are among those that have sought to engender pause and concern.[2] While varied in the terms and tones of their messages, each has raised a weighty question: Might the life sciences be the death of us?

The forewarning by Serguei Popov provides an illustrative example. As a leading scientist in the extensive Soviet biological-weapons programme until the early 1990s, he contributed to attempts to genetically enhance classic biowarfare agents as well as devise novel ones. Looking into the future on the basis of this past, in an article for *Technology Review* titled 'The Knowledge',[3]

1 For instance, see Rappert, B., Balmer, B. and Stone, J. 2008, 'Science, technology and the military: Priorities, preoccupations and possibilities', in *The handbook of science and technology studies*, London: MIT Press; James, A. 2007, 'Science & technology policy and international security', in Rappert B. (ed.), *Technology & security: Governing threats in the new millennium*, London: Palgrave.

2 For instance, see Lentzos, F. 2008, 'Countering misuse of life sciences through regulatory multiplicity', Science and Public Policy, vol. 35(1), pp. 55–64; Commission on the Intelligence Capabilities of the United States Regarding Weapons of Mass Destruction 2005, *Report of commission on the intelligence capabilities of the United States regarding weapons of mass destruction*, Washington, DC, available: http://govinfo.library.unt.edu/wmd/report/wmd_report.pdf [viewed 1 November 2009]; Fidler, D. and Gostin, L. 2008, *Biosecurity in a global age*, Stanford: Stanford University Press; Institute of Medicine and National Research Council 2006, *Globalization, biosecurity and the future of the life sciences*, Washington, DC: NRC; InterAcademy Panel 2005, *IAP statement on biosecurity*, 7 November, Trieste: IAP, available: http://www.nationalacademies.org/morenews/includes/IAP_Biosecurity.pdf [viewed 1 November 2009]; National Research Council 2004, *Biotechnology research in an age of terrorism*, Washington, DC: National Academies Press; NSABB 2007, *Proposed framework for the oversight of dual-use life sciences research*, Bethesda, MD: NSABB; World Health Organization 2006, *Biorisk management: Laboratory biosecurity guidance*, September, Geneva: WHO.

3 Williams, M. 2006, 'The Knowledge', *Technology Review*, March/April.

Popov outlined a number of accomplishments and possibilities in the Soviet programme that were becoming within reach of far-less-well resourced efforts. This was due to the growth in understanding of basic life processes and the accessibility of sophisticated technologies. Among the many prospects outlined included increasing the virulence of pathogens, synthesising viruses from common laboratory materials, modifying bacteria to induce debilitating diseases (including multiple sclerosis), as well as using pathogens to interfere with specific cellular targets in order to alter cognition, behaviour, and perception.

In the article, as typically happens elsewhere, such dire claims were accompanied by questions of a sceptical bent: would novel bioagents make for effective weapons in practice? Are sub-state groups or deranged individuals really in a position to produce them? Could the claims of former weapon developer be taken at face value? If it is relatively easy to deliberately spread disease, why have there not been more instances of bioattacks? Would controls on the conduct of research and the spread of technology make us safer or place us in greater danger?

'The Knowledge' concluded with a bleak assessment that: 'I don't know what kind of or scientific or political measures would guarantee that the new biology won't hurt us.' But the vital first step, Popov said, was for scientists to overcome their reluctance to discuss biological weapons. 'Public awareness is very important. I can't say it's a solution to this problem. Frankly, I don't see any solution right now. Yet first we have to be aware.'

Awareness and Education: Disagreement in Unanimity

Arguably these sentiments do not just represent the thinking of one man, but rather characterise the state of international thinking today regarding 'what must be done and by whom?'[4] As examined in the next section, a diverse array of assessments have been put forward about what dangers are associated with the life sciences and what should happen as a result. Calls for increased education of some kind have figured in recommendations across a range of concerns — from ensuring the physical safety of labs, to vetting experiment proposals, to tackling diseases that undermine economic development and thereby collective

4 See Rappert, B. and Gould, C. (eds) 2009, *Biosecurity: Origins, transformations and practices*, London: Palgrave.

well-being.[5] Education is envisioned as necessary in both those responses that call for informal self-governance by science communities as well as those that demand formal regulations.

However, as I have argued elsewhere, once one moves from such general calls to specific actions, a number of difficult choices must be addressed.[6] Initial indications of these are given in Box 1. It lists some basic questions that arise in considering what should be done.

Claims of what is appropriate biosecurity education are potentially contentious because they are bound up with the exercise of authority and expertise. For instance, with regard to concerns about purpose mentioned in Box 1, some types of education focus on transmitting authoritative knowledge or values. However, particularly in relation to matters of ethics, resistance can be intense when some try to tell others what they should think. Alternative types of education instead stress the need to nurture individuals' own reasoning so as to enable them to think through ethical problems on their own. Still other types are not focused on individuals, but seek to further the ability of people to work together in joint deliberations.[7] Not only are these different approaches associated with alternative learning techniques and opportunities for questioning, they also suggest various ways of resolving what should be done.

---

5 As a sample of such calls, see Report of Royal Society and Wellcome Trust Meeting 2004, 'Do no harm — Reducing the potential for the misuse of life-science research', 7 October; World Medical Association 2002, *Declaration of Washington on biological weapons*, Washington, DC: WMA; National Research Council 2003, *Biotechnology research in and age of terrorism*, Washington, DC: National Academies Press; British Medical Association 1999, *Biotechnology, weapons and humanity*, London: Harwood Academic Publishers; United Nations 2005, *Report of the meeting of States Parties to the Convention on the Prohibition of the Development, Production and Stockpiling of Bacteriological (Biological) and Toxin Weapons and on their Destruction*, BWC/MSP/2005/3 14, Geneva: UN, available: http://www.opbw.org [viewed 1 November 2009]; National Science Advisory Board for Biosecurity 2008, *Strategic plan for outreach and education on dual use research*, 10 December, available: http://oba.od.nih.gov/biosecurity/PDF/FinalNSABBReportonOutreachandEducationDec102008.pdf [viewed 1 November 2009].

6 Rappert, B. 2007a, 'Education for the life sciences' in Rappert, B. and McLeish, C. (eds) A Web of prevention: Biological weapons, life sciences and the future governance of research, London: Earthscan, pp. 51–65. Available: http://people.exeter.ac.uk/br201/Research/Publications/Chapter%203.pdf [viewed 1 November 2009].

7 See Päsänen, R. 2007, 'International education as an ethical issue' in Hayden, M., Levy, J. and Thompson, J. (eds) *Research in international education*, London: Sage, pp. 57–78.

Box 1: The Who, What, and How of Education

What should education entail by way of subject matter?

Should it include the characteristics of diseases from expected dangerous agents, the physical and biological security otf laboratories, the history of offensive programmes by states, or the potential of civilian research to further the spread of disease?

Who needs to be educated?

Should they be pathogen investigators, bioscientists as a whole, those associated with life sciences in general, or the public?

What is the purpose of education?

Should it seek to 'implant' knowledge or 'elicit' understanding?

Who is the educator?

In other words, who is expert?

How can audiences of practising scientists or other practitioners be reached?

How can their attention and active engagement be secured?

This volume examines a variety of attempts to bring greater awareness to security concerns associated with the life sciences. It identifies lessons from practical initiatives across a wide range of national contexts as well as more generic reflections about education and ethics. In offering their assessment about what must be done and by whom, each of the contributors addresses a host of challenging practical and conceptual questions. As a result, the volume will be of interest to those planning and undertaking activities elsewhere. In asking how education and ethics matter in an emerging area of unease, it will also be of interest to those with more general concerns about professional conduct and social problems.

Security and Biology: Dilemmas at Intersection

Before exploring the issues associated with education in more detail and introducing the chapters, this section continues focusing on the security implications of the life sciences. As will be argued, determining what to do by way of education is not only challenging because of the choices available in how to foster learning, but also because of the stubborn dilemmas, ambiguities and uncertainties associated with understanding the issues at stake. While it might seem plausible that growing knowledge and capabilities equate with growing agency for hostile actions, much scope for contest is also evident in conception of the issues at stake. The manner in which this is done has implications for what kind of education should be pursued, with whom, and how.

Consider, then, a number of contentious areas:

Running Faster

While in recent years some commentators have forwarded concerns about how developments in science and technology might aid the deliberate spread of disease, in practical terms, overwhelmingly research has been looked at as a way of countering identified threats. This is most marked in the US. Here a substantial expansion has taken place in biodefence and biodefence research. While in the financial year 2001 the US civilian biodefence funding totalled $569 million, in 2008 it was more than $5.3 billion.[8] Research has been a core component of this expansion, with funding in excess of $3 billion per year since 2004, much of it led by the National Institutes of Health.[9] In other words, the technologically sophisticated nightmares often envisioned have justified a similarly sophisticated response.

With the emphasis placed on staying ahead of threats through more research, worries have been expressed whether the shift in funding has established inappropriate priorities, blurred the boundary between internationally permissible defensive work, or created dangers regarding the accidental or intentional release of pathogens.[10] With regard to the latter, the substantial expansion of biodefence funding has resulted in a corresponding increase in the number of individuals and facilities working with pathogenic agents. Given the conclusion of the FBI that the perpetrator of the anthrax attacks in 2001 was an American working within the US Army Medical Research Institute of Infectious Diseases,[11] the question has been posed more than once as to whether the multi-billion-dollar increase in biodefence has resulted in a proliferation of dangerous knowledge, skills and materials. At the time of writing, intense

8 Franco, C. 2009, 'Billions for biodefence', *Biosecurity and Bioterrorism*, vol. 7(3): pp. 291–309.

9 See Center for Arms Control and Non-proliferation 2008, *Federal funding for biological weapons prevention and defense, Fiscal years 2001 to 2008*, Washington, DC: Center for Arms Control and Non-proliferation.

10 See Enserink, M. and Kaiser, J. 2005, 'Has biodefense gone overboard?', Science, vol. 307(5714), pp. 1396–98; Leitenberg, M., Leonard, J. and Spertzel, R. 2004, 'Biodefense crossing the line', *Politics and the Life Sciences*, vol. 22(2), pp. 1–2; Klotz, L. and Sylvester, E. 2009, *Breeding bioinsecurity*, Chicago: Chicago University Press.

11 Bhattacharjee, Y. 2009, 'The danger within', *Science*, 6 March, pp. 1282–83.

political debate is taking place in the US and elsewhere regarding what sort of screening and oversight of individuals should occur and who should control it. The underlying question of how much and what kind of defensive work is prudent remain topics on which governments and commentators have offered wildly opposing views.

In part, the question of what is appropriate defensive research is disputable because the ultimate ends served by that work are debatable. Much of the original increased funding in the US was designated for traditional 'Category A' agents (for example, anthrax, smallpox, plague). To some extent, in response to criticism about how this was establishing inappropriate research priorities, many of the funding programmes broadened their mission over time beyond a narrow conception of biodefence.[12] However, owing to the multiple dimensions in which goals can be mutually co-opted in the researcher–funder relation, determining the significance of official priorities has meant agendas and outcomes require a fine-grained analysis.

Everywhere and Nowhere

One aspect of increased research scrutiny that has animated much debate is the suggestion that work carried out in universities or other traditionally open organisations might be inappropriate to conduct or communicate. Unlike questions about the safety or physical security of labs, this does not so much relate to how research is conducted, but rather to its so-called dual-use dimensions. This usage of the term refers to the potential use of knowledge and techniques for beneficial and hostile purposes. Therefore, since 2003 a number of funders, publishers and organisations (in the West) have introduced oversight processes to assess the risks and benefits of individual instances of research to determine whether they need to be modified or withdrawn.[13]

It is notable that such procedures rarely conclude that manuscripts, grant applications or experiment proposals should not be undertaken or restricted. For instance, in 2003 a group of 32 science journals agreed general guidelines for modifying, and perhaps rejecting, manuscripts where 'the potential harm of publication outweighs the potential societal benefits'.[14] However, it would seem no manuscript has ever been rejected on security grounds.[15] As far as is known to the author, the same could be said of the funders that have established

12 See Franco 2009, op cit.

13 Rappert, B. 2008a, 'The benefits, risks, and threats of biotechnology', *Science & Public Policy*, vol. 35(1), pp. 37–44.

14 Journal Editors and Authors Group 2003, *PNAS*, vol. 100(4), p. 1464.

15 Van Aken, J. and Hunger, I. 2009, 'Biosecurity policies at international life-science journals', *Biosecurity and Bioterrorism*, vol. 7(1), pp. 61–72.

submission-oversight systems.[16] Perhaps even more notable with these review processes is the infrequency with which they have identified items 'of concern' in the first place.[17]

While data on research controls within government departments (especially defence-related ones) is not readily available, in relation to universities and other publicly funded agencies it seems justifiable to conclude that — barring dramatic changes — oversight processes will identify little research as posing security concerns and will stop next to nothing. This situation raises questions about the ultimate purposes and prospects of formal oversight procedures as well as who is conducting assessments and how (see below).

Formal Policies and Informal Practices

While many recently introduced formal dual-use procedures intended to weigh the perceived societal benefits and security risks of civilian research have not ruled any work should be halted, evidence suggests individual scientists might be acting otherwise. In 2007 the National Research Council and the American Association for the Advancement of Science (AAAS) conducted a survey of 10,000 AAAS members. The 2009 report of that survey indicated that one in six respondents had made some changes in what research they did, how it was communicated, or who it was done with.[18]

The small response size (16 per cent for completed surveys and 20 per cent for partially completed ones) means it is not possible to treat the findings as representative of any grouping. However, even without making generalised claims, disparities between the reported practices and the outcomes of recently introduced review processes are notable.[19]

While some have taken the survey findings to indicate scientists are already acting responsibly to reduce risks,[20] it seems more justified to ask further questions. One obvious question would be: why is there such inconsistency between the willingness of researchers to report forgoing aspects of their work with dual-use potential and the inability of formal process to do the same (or

16 These include the UK Biotechnology and Biological Sciences Research Council, the UK Medical Research Council, the Wellcome Trust, the Center for Disease Control, and the Southeast Center of Regional Excellence for Emerging Infectious Diseases and Biodefence.

17 So across all the journals in the Nature Publishing Group, roughly 15 papers were subjected to a special security review in 2005 and 2006. For further figures, see Rappert 2008a, op cit.

18 National Research Council and the American Association for the Advancement of Science 2009, *A survey of attitudes and actions on dual-use research in the life sciences*, Washington, DC: NRC and AAAS.

19 Though in this regard, 25 per cent of respondents to the survey indicated they had worked with 'select agents' in the past, therefore suggesting a possible reason for both the changes made to research practices and the awareness of dual-use concerns.

20 National Academies 2009, *Survey samples life scientists' views on 'dual use' research and bioterrorism*, Press Release 9 February.

even find any aspects of concern)? Within organisational sociology, the disparity between formal procedures and informal practices is a long-standing topic of commentary. Such a situation is not necessarily an occasion for arguing who has made the right decisions, but rather one for asking how alternative assumptions about proper working practices are informing conduct.

In relation to the themes of education central to this book, the disparity also raises the issue of how practitioners communicate concerns. Particularly in light of the lack of dual-use educational provisions as part of university degree programmes and the absence of professional attention to this topic in recent decades, how the scientists surveyed became concerned by hostile applications is an important matter that could signal pathways for educational interventions, such as seeking to make explicit practices that were previously implicit.[21]

Individual versus Cumulative Developments

It seems reasonable to argue that one reason why dual-use risk-benefit review processes have not halted grant applications, manuscripts, and experiment applications is the difficulty of establishing the possible hazards associated with single-research inputs against the backdrop of pre-existing knowledge and capabilities. Despite ongoing attempts,[22] making risk determinations is highly problematic. Even if reasonably robust assessment procedures could be devised, it is not clear that threats derive from discrete projects, so much as how cumulative developments in knowledge, know-how, and technologies enable additional possibilities for action. Just how that is happening is essential to understanding what is possible (see below).

Therefore, rather than focusing on whether particular experiments should go ahead, it seems more fruitful to ask what directions should be funded in the first place.[23] Some of the lines of biodefence undertaken in the US and elsewhere (particularly those associated with characterising threats) might be questionable in terms of their necessity. Positively, directions of work that might enhance security by fostering international collaboration and development, as suggested in the *DNA for Peace* initiative, could be supported.[24] This 'macro' attention towards research directions is not without problems too, such as how

---

21 For a discussion of this as a prevalent form of ethical training, see Halpren, S. 2004, *Lesser harms*, Chicago: Chicago University Press.

22 Royal Society 2009, *New approaches to biological risk assessment*,29 July, London: Royal Society.

23 Johnson, D. 1999, 'Reframing the question of forbidden knowledge for modern science', *Science and Engineering Ethics*, vol. 5(4), pp. 445–61.

24 *DNA for peace: Reconciling biodevelopment and biosecurity*,available: http://www.utoronto.ca/jcb/home/documents/DNA_Peace.pdf [viewed 1 November 2009].

to anticipate research results. However, what is clear is that attention to date on dual-use issues has been directed at individual elements of research at the expense of other approaches.

'Dual Use' is... 'X' is...

In the paragraphs above, 'dual use' refers to the potential for knowledge and techniques to serve beneficial and hostile purposes. In doing so, it is roughly in line with the highly influential report by the US National Academies, *Biotechnology Research in an Age of Terrorism*.[25] However, like other terms with a rising currency of late — such as 'biosecurity'[26] or 'codes of conduct'[27] — 'dual use' has its own history. It is often taken to mean different things by different people. That ambiguity has no doubt been part of its attraction.

This indistinctness brings certain hazards too. One is a lack of clarity and corresponding misunderstanding in what is being argued about the relation between science and security. For instance, Atlas and Dando have offered a distinction between three dual-use aspects of the life sciences to avoid conflations. They distinguish between: 1) how notionally civilian facilities can be used to develop biological weapons; 2) how agents and equipment intended for peaceful purposes can be used in the production of bioweapons; and 3) how knowledge generated through science can aid those seeking to produce weapons.[28] They argue each is associated with its own conundrums and require specific types of responses — mandatory international inspections and transparency in the case of facilities; balanced export controls and domestic oversight in the case of agents and equipment; and a culture of responsibility in relation to knowledge.

Other matters are at stake in our use of terminology and concepts than possible misunderstanding. As McLeish argues, questionable assumptions can often underlie reference to the dual-use presumptions that rarely get scrutinised because of the ready labelling of science and technology as such. As she argues, much of the security analysis relies on an outdated linear model of innovation wherein science is applied to produce new technologies. In addition, the locus of concern with dual-use issues in many commentaries often shifts in an uneasy and unacknowledged manner between the transfer of materials, the intention of users of knowledge and technology, and the physical characteristics of technology itself.

---

25 National Research Council 2004, *Biotechnology research in an age of terrorism*, Washington, DC: National Academies Press.

26 See Rappert and Gould (eds) 2009, op. cit.

27 Rappert, B. 2007b, 'Codes of conduct and biological weapons', *Biosecurity & Bioterrorism*, vol. 5(2), pp. 145–54.

28 Atlas, R. and Dando, M. 2006, 'The dual-use dilemma for the life sciences: Perspectives, conundrums, and global solutions', *Biosecurity and Bioterrorism*, vol. 4(3), pp. 1–11.

For such reasons, the language adopted to characterise the security dimensions of science and technology can also cloud understanding.[29]

What are Biological Weapons?

Greatly aiding efforts to prevent the deliberate spread of disease is the widespread denunciation of any such act. As opposed to other fields, such as nuclear science, weaponising the latest research findings is generally seen as inappropriate in the life sciences. Within customary and international law, as well as the rhetoric of states, 'biological weapons' are set apart from other weapons in that they are treated as a distinct category.[30]

In some ways this has only displaced controversy to the question of what counts as a biological weapon in the first place.[31] This is most evident in debates about the appropriateness of biochemical compounds as instruments of force.[32] The use by Russian security forces of a fentenyl gas (an opium-based narcotic) during the Moscow theatre siege in 2002 provides one example of the types of options being pursued by states for law enforcement and military operations that might be designated 'biological'. Additionally, governments such as the US have examined more sophisticated biochemical choices to alter consciousness, behaviour and emotions.[33] The acceptability and permissibility of such biochemical agents is fought out, in part, through terminology. Proponents make use of labels such as 'calmatives', 'incapacitants' and (misleadingly) 'non-lethal weapons'.[34] Should such developments lead to a legitimate role for bioagents as a means of force, the implications for the current stigmatisation of biological weapons would likely be substantial.

What are 'Effective' Biological Weapons?

As a final area of contention, much disagreement is evident today regarding the extent of biothreats. Some of this stems from underlining presumptions about

---

29 It should be kept in mind though that ambiguity in meaning is often highly valuable in building shared agendas.

30 This in contrast to the suggestion by many of those involved in offensive programmes that biological weapons are not different from others. See Domaradskil, I. and Orent, W. 2003, *Biowarrior: Inside the Soviet/Russian biological war machine,* Amherst, NY: Prometheus, p. 150; Balmer, B. 2002, 'Killing "without" the distressing preliminaries', *Minerva,* vol. 40(1), pp. 57–75.

31 See Rappert, B. 2006, *Controlling the weapons of war: Politics, persuasion and the prohibition of inhumanity,* London: Routledge, Chapter 6.

32 For an overview, see Pearson, A., Chevrier, M. and Wheelis, M. (eds) 2007, *Incapacitating biochemical weapons: Promise or peril?,* Lanham, MA: Lexington Books.

33 Dando, M. 2009, 'Biologists napping while work militarized', *Nature,* vol. 460, p. 950; British Medical Association 2007, *Drugs as weapons,* London: BMA House.

34 See Rappert, B. 2003, *Non-lethal weapons as legitimizing forces?: Technology, politics and the management of conflict,* London: Frank Cass.

what counts as a concern. Certainly within the West, much emphasis is with sub-state groups. The limited number of bioterror attacks in the past and the difficulties experienced by even well-funded groups using classic pathogens (for instance, the Japanese Aum Shinrikyo cult[35]) suggest a low likelihood of mass casualties by terrorist groups acting alone.[36] Therefore, the possibility that such groups could or would want to make use of today's cutting-edge science is even more remote, at least anytime 'soon'.

However, the situation is more complex than this. Even if it is accepted that inflicting mass casualties requires a well-resourced state programme, concerns can derive from the fear and disruption caused by deliberate spread of disease. As illustrated in the case of the 2001 anthrax letters, attacks need not inflict mass casualties to be highly consequential. 'Weapons of mass disruption' rather than 'weapons of mass destruction' sums up a contrast.

Concern can intensify when the basis for disruption is analysed. Fundamental to the international prohibition of bioweapons today is the view that these weapons are especially abhorrent. That orientation is expressed in international accords such as the Biological Weapons Convention. The continuing promotion of such agreements and related rules, in turn, reinforces this negative standing. Indeed, it is the manner in which biological weapons are treated as distinctly repugnant that would likely contribute to significant fear and disruption in the case of an attack.

Education as...

The previous section posed some major weaknesses in thinking about the life sciences–security relation. When this is understood to involve uncertainties and unknowns where much scope exists for disagreement, the question of what should happen by way of education becomes less straightforward than it might initially appear.

This section adds further density to the picture. While some of the who's, what's, and how's of education were noted above, this section examines the multiple roles, functions, and standing sought for education. The goal is not to consider the details of what teaching efforts should include, but how education in general

35 See Furukawa, K. 2009, 'Dealing with the dual-use aspects of life science activities in Japan', in Rappert and Gould (eds), op. cit.

36 As argued in Ouagrham-Gormley, S and Vogel, K. 2010, 'The social context shaping bioweapons (non) proliferation', *Biosecurity & Bioterrorism*, Volume 8(1) and Leitenberg, M. 2001, *Biological weapons in the twentieth century: A review and analysis*, Washington, DC: FAS, available: http://www.fas.org/bwc/papers/bw20th.htm [viewed 1 November 2009]. For a critical response to the claims in this chapter by Popov, see Macfarlane, A. 2006, 'Assessing the threat', *Technology Review*, March/April.

is positioned within debates about the science–security relation. By considering the many things education can be understood as, this section will help position the practical initiatives and proposals summarised in the next section. As will be apparent, the issues at stake extend well beyond what individuals sitting at benches know or think.

...Prerequisite

Education can be treated as necessary for other security-related activities to be undertaken. For instance, much store has been placed in professional codes of conduct since 2001.[37] Such options have been said by many to be a way of promoting self-governance. The circulation of codes would foster a culture of responsibility by making scientists more aware and providing ethical guidance. However, efforts to devise meaningful codes have largely floundered. In no small part, this has been due to the lack of prior awareness and attention by researchers as well as science organisations to the destructive applications of the life sciences.[38] Before codes can help teach, education is needed.

Also, consider the dual-use reviews noted in the previous section. The Wellcome Trust, the British Biological Sciences Research Council and the British Medical Research Council are among those funders that have established grant-review procedures. Each relies on applicants to self-identify cases where work could generate outcomes open to misuse for harmful purposes. In light of the limited professional attention to this possibility in recent decades, it seems quite likely that a lower identification rate is taking place than would be the case with a highly dual-use-aware community of applicants.

The US National Science Advisory Board for Biosecurity (NSABB) was established to advise the federal government how to respond to the dual-use potential of the life sciences. It similarly advocates a system for the oversight of experiments that relies on lead investigators to undertake the initial determination of whether their work is 'of concern'.[39] In recognition of the need for those making such assessments to be cognisant of security threats though, as well as other provisions it recommends that: 'All federal agencies involved in the conduct and support of life sciences research [...] should require that their employees, contractors, and institutional grantees train all research staff in the identification and management of dual-use research of concern.'[40]

---

37 Rappert, B. 2009, *Experimental secrets: International security, codes, and the future of research*, New York: University Press of America.

38 Rappert 2007b, op. cit.

39 NSABB 2007, *Proposed framework for the oversight of dual-use life sciences research: Strategies for minimizing the potential misuse of research information*, Bethesda, MD: NSABB.

40 NSABB 2008, *Strategic plan for outreach and education on dual-use research issues*, Bethesda, MD: NSABB.

Some have argued this goes far enough. An alternative model for oversight proposed by a group at the University of Maryland advocates independent peer reviewers should carry out the identification of what is of concern. In part, this was justified by citing the likely limits on the security expertise of researchers, even if they had undergone some formal training.[41]

...Deficiency Correcting

Much of the current analysis of what practising researchers know — be that regarding laboratory physical security[42] or wider ethical/arms-control issues — posits a deficiency model.[43] That is to say, they note a lack of knowledge held by certain groups. Education is advocated as a way to correct that ignorance. Knowledge is taken to be good: more knowledge leads to better decisions. Depending on the size of the hole perceived and the value attached to additional knowledge, a call is made for voluntary or mandatory measures, often through formal teaching.[44]

The particulars of how deficiency is portrayed are highly consequential in framing what kinds of problems exist and how they can be addressed. Take the case of the NSABB Working Group on Communication.[45] In line with the review processes adopted by certain journals, the NSABB Charter required it to 'advise on national policies governing publication, public communication, and dissemination of dual-use research methodologies and results'. At the first public meeting of the NSABB in late 2005, the Communication Working Group stated it would:

> - Identify concerns and examine options and strategies for addressing issues related to the communication of dual-use research information.
>
> - Develop draft recommendations for the NSABB that will facilitate the consistent application of well-considered principles to decisions about communication of information with biosecurity implications.[46]

41 Harris, E. 2007, 'Dual-use biotechnology research: The case for protective oversight', in Rappert and McLeish (eds) op. cit.

42 For a range of analyses of what is known by researchers around the world on this matter, see the publications of the International Biological Threat Reduction Group and Sandia National Laboratories at http://www.biosecurity.sandia.gov/main.html?subpages/documents.html.

43 For a further discussion of this model within discussions about science, see Bush, J., Moffatt, S. and Dunn, C. 2001, 'Keeping the public informed?', Public Understanding of Science, vol. 10, pp. 213–29.

44 For an instance of the latter, see Rappert, B. and Davidson, M. 2008, 'Improving oversight: development of an educational module on dual-use research in the West', Conference Proceeding for Promoting Biosafety and Biosecurity within the Life Sciences: An International Workshop in East Africa, 11 March, Kampala: Ugandan Academy of Sciences.

45 For a further analysis of this example, see Rappert, B. 2008b, 'Defining the emerging concern with biosecurity', Japan Journal for Science, Technology and Society, vol. 17, pp. 95–116.

46 Kiem, P. 2005, 'Working group on communication of dual-use research results, methods, and technologies', Meeting of National Science Advisory Board for Biosecurity, 21 November, Bethesda, MD.

As such, 'the problem' was to reduce the prospect that otherwise benignly intended findings might aid development of bioweapons. This framing shifted quite quickly. By July 2006 'the public' assumed a prominent position within deliberations of the working group. Because of an acknowledged lack of understanding of science, the fear repeatedly expressed was that future media reports could spur the public to demand (inappropriate) dissemination restrictions. In response, the Communication Working Group's main charges deriving from the NSABB Charter were modified to:

- Facilitate consistent and well-considered decisions about communication of information with biosecurity implications.
- Demonstrate to the public that scientists recognise, and are being responsive to, concerns about the security implications of their work.[47]

With the latter requiring the public to be properly informed about the dangers posed from open publishing (manageable, relatively limited, etc.). This re-specification was in line with a wider movement within the NSABB deliberations to focus on the 'threats *from* science' (and thus the need for new polices and oversight measures) while also considering the 'threats *to* science' (from new polices and oversight measures). In this way, alternative notions of who is deficient suggest other problems and solutions.

Following on from this, more knowledge is not always seen as good, at least not entirely. While it might regularly be advocated that researchers should be more cognisant of the dual-use potential of science and technology, the same cannot be said of 'the public'. Much debate is evident about just how loudly security concerns should be made known to the population at large.[48] Scant efforts made prior to 2001 (and even since) by scientists to popularise how their work might aid the production of bioweapons indicate the historical pattern of *not* seeking to foster wider debate and awareness. So while many have dismissed the security concerns associated with the publication of certain experiments — such as the IL-4 mousepox and the synthetic creation of poliovirus — because their findings were already well known among specialists,[49] the question can be asked why such possibilities were not more widely mooted before.

This last point also raises the prospect that without more engagement from practising researchers, policy and security analysts might be forming inappropriate threat assessments because of the haphazard way certain concerns have received a wide airing.

---

47 Kiem, P. 2006, 'Working group on communication of dual-use research results, methods, and technologies', Meeting of National Science Advisory Board for Biosecurity, 13 July, Bethesda, MD.

48 For a discussion of this, see Rappert, B. 2007c, *Biotechnology, security and the search for limits: An inquiry into research and methods*, London: Palgrave, Chapter 5.

49 As in Block, S. 2002, 'A Not-so-cheap Stunt', *Science*, vol. 297(5582), pp. 769–70.

...Irrelevant

Another way to question the treatment of education is simply to question its utility. A fundamental tenet in social research is that behaviour is more influenced by situational conditions than personal dispositions. In other words, what people do is highly dependent on the situations in which they find themselves and the pressures they experience. As such, it is little wonder so many doctors, biologists, engineers and others in the Soviet Union were willing to take part in its offensive programme. While it is unclear how many involved knew about the international prohibition enshrined in the Biological Weapons Convention, it does seem clear that familiarity with the text would have done little to alter their participation.

Besides such coercive environments, it can be said that standards of ethics and practice are intertwined with the imperatives under which individuals operate. For instance, how those in hospitals deal with the confounding choices about life and death experienced on a daily basis must be understood as being indebted to the social organisation of the hospital itself.[50] In these settings, mundane issues such as the relative power of nurses, doctors and administrators are highly pertinent in what decisions are made. Therefore, it would not be enough for nurses to receive instruction about ethical principles for them to act in a way they would regard as right. Similarly, in the case of laboratory researchers, the cultures and reward structures of labs (for example, 'publish or perish') can reduce the prospects of individuals acting in a way they judge as proper, or can work against the recognition of ethical problem in the first place.[51] This can take place whatever individuals believe should be the case.

Thus in any discussion about education, it is necessary to ask where it can be made to matter and what weight it can be expected to bear.

...a Social Problem

While the extensive participation of experts in offensive bioweapons programmes during the twentieth century suggests limitations to education, this history also indicates the potential for it to contribute to hostile activities. When ethics instruction is conceived as the dissemination of values, just what those values are is central in evaluating the benefits of what is learnt. If duty to

50 Chambliss, D. 1996, *Beyond caring: Hospitals, nurses and the social organization of ethics*, London: University of Chicago Press.

51 National Research Council 2009, *Ethics education and scientific and engineering research: What's been learned? What should be done?*, Washington, DC: NRC, Chapter 6. For a wider examination of this, see Vaughan, D. 1996, *The Challenger launch decision*, Chicago: Chicago University Press.

one's country[52] is taught to be the paramount concern, then it easy to envision how education might not always be a 'force for good'.[53] The method of education can foster relations of subordination too that are contrary to maintaining peace.

More widely, efforts to instruct always come with commitments and assumptions. These can mean that education functions to maintain existing inequalities.[54] So while the unacceptability of biological weapons is generally unquestioned within diplomatic circles today, that does not necessarily make for a non-problematic common pedagogical message. As Richard Falk suggested, the recent regime restricting bioweapons must be seen in its political context. Writing in 2001, he suggested this context was one in which the US was trying to divert public attention away from existing US nuclear capabilities. It was against this selective prioritisation that he asked whether:

> The ongoing process that supports CW [chemical warfare] and BW [biological warfare] regimes, as well as the nuclear non-proliferation treaty regimes, [should] be re-evaluated and possibly rejected? From the perspective of the equality of states, a fundamental norm in international law, are these regimes embodiments of the hegemonic structure of world politics that controls and deforms diplomatic practice?[55]

Herein, what is good and why must always be understood 'in context', though what counts as the right context is the stuff of political debate. Any education message is going to contest competing notions about what is right. When approached in this way, it is not hard to see how others could interpret efforts taken in one country to promote security mindedness as imperialistic.

Somewhat less critical, certain types of education can be thought to take time and energy away from dealing with causes of problems. For instance, abstract and hypothetical instruction about ethical problem-solving that is removed from the real experiences of individuals can do more harm than good. To the extent that ethics is taught without reference to power relations that give rise to practical conflicts and dilemmas, it can mask the sources of tension and perpetuate inaction.[56]

---

52 Or group difference, see Nelles, W. (ed.) 2004, *Comparative education, terrorism and human security*, London: Palgrave.

53 Moving outside a consideration of biological weapons, it can be noted that the education system in a wide range of countries does little to dissuade individuals from using their knowledge and skills to perfect forms of killing.

54 Saltman, K. and Gabbard, D. (eds) 2003, *Education as enforcement: the militarization and corporatization of schools*, London: Routledge; Apple, M. 2000, *Official knowledge: democratic knowledge in a conservative age*, London: Routledge; and Harber, C. 2004, *Schooling as violence: how schools harm pupils and societies*, London: Routledge.

55 Falk, R. 2001, 'The challenges of biological weaponry', in Wright, S. (ed.), *Biological warfare and disarmament*, London: Rowman & Littlefield, p. 29.

56 For an analysis of this, see Chambliss 1996, op cit., Chapter 4.

...a Diversion

Consider another concern, that of treating education as distraction. This does not pertain to those being educated, but to those talking about the need to educate others. For instance, in recent years the awareness and training of scientists has been a topic of international consideration within the States Parties meetings of the Biological and Toxin Weapons Convention (BTWC). Since the failure in 2001–02 to agree a legally binding verification measure for the convention, yearly meetings have been structured through an intersessional process. In this, governments engage in non-binding discussions about selected topics. The education of scientists figured as a significant theme within the 2003, 2005 and 2008 meetings.

However, from its inception, questions have been raised about the value of the intersessional process. At least one reason is that it distracts delegates from attempts to agree verification instruments or other compulsory measures.[57] Worse still, over time it might establish low expectations. Even if one concludes the BTWC intersessional process has been useful, how long 'mere' discussion should go on is open to debate. Comparing biosecurity education against other possibilities for action (or even more generic education possibilities) shows how the worth of any activity can be queried.

...Guardianship

Alternatively, and more positively, the education of scientists could be seen as a way of ensuring conventions and agreements — such as the BTWC — remain meaningful. This is because further awareness of the security implications of science and of the international instruments for the prohibition of bioweapons leads to more engaged scientific communities. By taking greater notice of and participation in the relevant activities, practitioners could help ensure governments are aware of their own commitments and labour to undertake effectual actions. By extending the range of those working to eliminate bioweapons, this could also reinforce the stigma against the deliberate spread of disease.[58]

---

57 See Chevrier, M. 2002/03, 'Waiting for Godot or saving the show? The BWC Review Conference reaches modest agreement', *Disarmament Diplomacy*, No. 68, available: http://www.acronym.org.uk/dd/dd68/68bwc.htm [viewed 1 November 2009].

58 For a further argument of these points, see Revill, J. and Dando, M. 2008, 'Life scientists and the need for a culture of responsibility: After education...what?' *Science and Public Policy*, vol. 35(1), pp. 29–36.

...Catalyst

Additionally, the widespread concord regarding the value of awareness and education could be utilised as a stepping-stone for moving ahead on other matters. Within the recent meetings of the BTWC a number of issues, such as promoting international co-operation and verification measures, have proven highly contentious. Likewise, proposals for oversight and dual-use reviews of research can generate heated debate. However, awareness and education are matters on which those from security, scientific, diplomatic, and other backgrounds can reach general consensus. Therefore, achieving understanding and progress here could help secure advancement in relation to more overtly problematic areas. It might also help indicate how to establish progress. In relation to the BTWC, the discussion-only terms set for the intersessional process since 2003 have led to an absence of the international targets and metrics. As a comparatively approachable topic, setting international standards for education could be agreed as a way of opening peoples' imaginations to the possibility of setting other measurable goals.[59]

...Enrolment

The last two sub-sections, in particular, questioned whether education is orientated towards being an end itself or a preliminary step towards another, secondary, end. As has happened in relation to codes of conduct, contrasting assumptions can underlie similar calls for action. In the case of codes, those (often implicit) assumptions related to whether their adoption would placate the need for additional oversight measures or whether they were part of a stepwise movement towards comprehensive systems of control (for instance, the licensing of scientists).[60]

This suggests the need to interpret calls for education as part of enrolment processes. Through setting agendas, framing problems, and establishing interested networks, what is being done today is helping to form possibilities for future action. Whether through purposeful direction or unintended preoccupation, the choices made about what kind of education should be pursued or how it is being discussed are shaping directions for the future. Just how much this is taking place and in which directions are important considerations.

---

59 For a discussion of possible international goals, see Dando, M. 2008, 'Acting to educate life scientists', *Bulletin of the Atomic Scientists*, 31 October, available: http://www.thebulletin.org/web-edition/columnists/malcolm-dando/acting-to-educate-life-scientists [viewed 1 November 2009].

60 Rappert 2009, op cit.

The Chapters

The previous section suggested the range of possibilities that can be sought from education as well as talk of education. The remaining chapters demonstrate that variety by describing diverse efforts to educate scientists and others about the life science–security relation.

The chapters in Part One begin by extending the analysis of education presented in this introduction to questions of ethics. Selgelid examines the intersection of ethics and dual-use concerns. His goal is not only to outline the ethical dimensions of the multiple uses of research, but also to ask why bioethics has had so little to say until recently about this topic. In Chapter 2, Sture analyses what lessons past developments in medical and business ethics hold.

Part Two recounts national experiences to promote awareness and institute educational measures. In doing so, the chapters detail the current attention paid to biosecurity and dual-use issues in the countries under consideration. As will be evident, different countries are in very varied situations with regards to their past engagement and to their basis for moving forward. In describing their experiences, Garraux, Friedman, Minehata and Shinomiya, Barr and Zhang, Connell and McCluskey, as well as Enemark, provide many entry points and models for promoting education.

Part Three moves on from national activities to reflect on international possibilities. Mancini and Revill review their efforts to establish a collaborative Biosecurity Education Network. Part of that entailed the presentation of tailored educational material to life-science students. In Chapter 10, Whitby and Dando consider the rationale for the Biosecurity Education Module Resource noted by Mancini and Revill, including how it could figure within the work of civil organisations as well as the BTWC. Johnson then asks how ethics training about security issues could be made professionally relevant for scientists, in particular, by advocating the potential of role-playing exercises.[61] In the Conclusion, Bezuidenhout and I draw together strains from these chapters in an effort to point the way for future action and research.

61 For a further consideration of role-playing exercises for biological weapons, see Rappert, B., Chevrier, M. and Dando, M. 2006, In-Depth Implementation Publications of the BTWC: Education and Outreach Bradford Review Conference Paper No. 18, available: http://www.brad.ac.uk/acad/sbtwc/briefing/RCP_18.pdf [viewed 1 November 2009].

Chapter 1: Ethics Engagement of the Dual-Use Dilemma: Progress and Potential

MICHAEL J. SELGELID

Introduction

During the past decade, the problem of dual-use research, science, and technology has been one of the most debated issues in discourse surrounding biological weapons and the bioterrorist threat, and a particularly controversial topic regarding science policy. The expression 'dual use' was historically used to refer to technology, equipment, and facilities that could be used for both civilian and military purposes. Conceived this way, dual-use technology is not necessarily something to worry about. To the contrary, this kind of technology was sometimes considered desirable from the standpoint of policymakers — a way of killing two birds with one stone. Policymakers were nonetheless concerned about exporting such technologies to adversary countries.

In contemporary discourse the expression 'dual use' is usually used to refer to research, science and technology that can be used for both good and bad purposes. While almost anything can have multiple functions, current debates have been primarily concerned with bad purposes involving weapons — and, most commonly, weapons of mass destruction in particular (that is, where the consequences of malevolent use would be most severe). Of specific concern is the possibility that recent developments in the life sciences may enable development of a new generation of especially dangerous bioweapons.[1]

Such concerns are illustrated by a number of controversial experiments published during the past decade, such as the genetic engineering of a superstrain of vaccine-resistant mousepox,[2] the artificial synthesis of a live polio

1 National Research Council 2004, *Biotechnology research in an age of terrorism*, Washington, DC: National Academies Press.

2 Jackson, R. J., Ramsay, A. J., Christensen, C. D., Beaton, S., Hull, D. F. and Ramshaw, I. A. 2001, 'Expression of mouse interleukin-4 by a recombinant ectromelia virus suppresses cytolytic lymphocyte responses and overcomes genetic resistance to mousepox', *Journal of Virology*, vol. 75, pp. 1205–10.

virus from scratch,[3] and the reconstruction of the 1918 Spanish Flu virus;[4] and the general phenomenon of converging technologies, such as synthetic biology and bionanotechnology.[5] Given these and other developments, life scientists are currently in a situation very similar to that faced by atomic physicists early in the twentieth century, when key discoveries that enabled production (and use) of the first atomic bombs were made. Like nuclear technology, powerful technologies made possible by the rapid progress of the life sciences may have great benefits for humankind, but they could also have disastrous consequences if employed by those bent on causing destruction.

The dual-use phenomenon raises important questions about the responsibilities of scientists, research institutions, the scientific community, publishers, and policymakers. Responsible actors at each of these levels should aim to promote the progress of science insofar as such progress will benefit humanity; but they should aim to avoid outcomes where developments ultimately result in more harm than good. One popular idea in recent debates about the dual-use problem is that we should aim for policy that strikes a balance between the goal to promote scientific progress (and the good things thereby enabled) and the goal to protect security.[6]

While that sounds plausible, open questions remain: What, for example, would be an appropriate balance between scientific progress and security; and how could such a balance be attained in practice? Should we rely on voluntary self-governance, or is more governmental oversight called for? Perhaps unsurprisingly, the scientific community is strongly in favour of 'bottom-up' solutions to dual-use research governance (that is, voluntary self-governance). It is commonly held that new codes of conduct addressing responsibilities related to dual-use research should be adopted, and that scientists should be further educated about the potential dual-use implications of their work; but scientists generally resist the idea that solutions to problems raised by the dual-use phenomenon should involve increased governmental control over scientific enterprise. Among other things, they argue that autonomy is essential to science progress and that governmental interference would be both counterproductive (unnecessarily stifling important and beneficial research) and violate the right to freedom of inquiry (academic freedom) and (in the case of governmental censorship of dangerous discoveries) freedom of speech.

---

3 Cello, J., Paul, A. V. and Wimmer, E. 2002, 'Chemical synthesis of poliovirus cDNA: Generation of an infectious virus in the absence of natural template', *Science*, vol. 9, pp. 1016–18.

4 Tumpey, T. M., Basler, C. F., Aguilar, P. V., Zeng, H., Solórzano, A., Swayne, D. E., Cox, N. J., Katz, J. M., Taubenberger, J. K., Palese, P. and García-Sastre, A. 2005, 'Characterization of the reconstructed 1918 Spanish influenza pandemic virus', *Science*, vol. 310, pp. 77–80.

5 Institute of Medicine and National Research Council 2006, *Globalization, biosecurity and the future of the life sciences*, Washington, DC: National Academies Press.

6 National Research Council 2004, op cit.

Bioethics' Neglect of Dual-Use Discourse

The main parties to debates about the responsibilities of scientists and other actors — and relevant debates about the governance of dual-use research — have, to date, mainly been scientists and security experts. With the exception of a small and recently emerging literature, notably absent from such debates has been the voice of (bio)ethicists in particular. This is unfortunate partly because the dual-use dilemma is, by its very nature, an ethical one. Talk about 'beneficial and malevolent' or 'good and bad' uses of science and technology, the 'promotion of benefits and the avoidance of harms', and the 'responsibilities' of actors all fall squarely within the realm of ethics: the discipline explicitly concerned with issues of good and bad, right and wrong, and the duties and responsibilities of human beings.

Above, I indicated that a plausible and popular notion is that we should aim for policy that strikes a balance between the goal to promote security and the goal to promote scientific progress; the idea being that heavy regulation of science might promote security at too high a price with regard to scientific progress and that too little oversight might facilitate science progress at too high a price with regard to security. This raises questions about the (nature of the) value of security and the (nature of the) value of scientific progress — and questions about how such values should be balanced against one another in cases of conflict. For example, should security be considered to be merely of instrumental value — something that should be valued only insofar as it promotes (other) things that are considered to be intrinsically valuable (that is, valued for their own sake)? Or is security itself intrinsically valuable? These are all ethical questions.

Questions about norms, values and what social policy should be are precisely the kinds of things that ethics is concerned with. While science is concerned with what is the case; ethics, by definition, is concerned with what should or ought to be the case. Although their contribution to debates about dual use is (for obvious reasons) absolutely essential, scientists and security experts have no special expertise for analysis of normative questions such as these, and so more input from ethicists is crucial. A broader, more interdisciplinary discussion about the dual-use problem is wanted.

Given that bioethicists have had so much to say about research ethics in general and the social implications of genetic research and science in particular, it is surprising that they have not been more actively engaged in discussions about dual-use life-science research. Discourse surrounding research ethics has traditionally focused on the protection of human and animal subjects rather than dangers associated with the potential malevolent use of research findings. Recent decades have witnessed enormous attention from bioethicists to 'ethical, legal, and social implications' (ELSI) of the genetics revolution. In the early days of DNA recombinant technology, environmental safety was a focus of bioethics discussion regarding genetics research. More recently ELSI discourse surrounding genetics has focused on the (clinical) safety of genetic therapy (or research pertaining thereto), genetic determinism, genetic testing, genetic discrimination, genetic enhancement, selective reproduction (that is, eugenics), cloning, stem-cell research, DNA fingerprinting, and the patenting of genetic sequences. That these have, to date, undeniably been the standard topics of ELSI discourse is quickly revealed by examination of titles, abstracts, tables of contents, and indexes of texts concerned with ELSI issues surrounding genetics. Though concern regarding dual-use life-science research largely relates to genetics in particular, and although the weapons implications of genetics may turn out to be the most serious (ethical and social) consequence of the genetics revolution, those concerned with the ethical implications of genetics have traditionally (and until only very recently, and in exceptional cases) been almost entirely silent about the potential weapons implications of genetics.

The lack of bioethics' attention to this topic is partly revealed by Robert Cooke-Deegan's canonical history of the Human Genome Project, *The Gene Wars.*[7] Cooke-Deegan's volume explicitly includes coverage of the politics and ethical debate surrounding the new genetics, and it even includes a chapter entitled 'Genes and the Bomb'. Despite the links it draws between genetics and atomic weapons, however, the volume never mentions the biological-weapons implications of genetics. This is odd partly because (as Cooke-Deegan demonstrates) important origins of the Human Genome Project are found in the US Department of Energy and the Los Alamos laboratories where the first atomic bombs were made. Such organisations were interested in genetics partly because they wanted to learn about radiation's effects on genetic material. Given these organisations' explicit concern with (albeit nuclear) weapons of mass destruction, not to mention their governmental and military affiliations, one expects that those involved would have recognised and considered the weapons potential of the genetics revolution very early on. Therefore, it is surprising that discussion of such issues is not included in Cooke-Deegan's commentary on debates surrounding socially controversial aspects of genetics.

It is commonly said in ELSI discourse that the power of genetics is comparable to the power of atomic physics, and that we thus need more ethical discussion and reflection about the former than the latter received when the first atomic bombs were made and used — so that more socially responsible decisions about science can be made in genetics than were made regarding nuclear energy. However, the usual topics of ELSI discourse reveal that weapons development is not what

7 Cooke-Deegan, R. 1994, *The gene wars: Science, politics, and the human genome*, New York: Norton. The discussion that follows is not meant to be a critique of Cooke-Deegan.

those concerned with the ethics of genetics have had in mind. At the time of writing (in early 2010), an enormous number of journal articles and books on ethics and genetics have been written; but, aside from a few recent exceptions, explicitly ethical literature (authored by ethicists and in books and journals primarily concerned with ethics) includes little if any discussion of genetics' potential role in weapons making.

This raises interesting historical and sociological questions about the discipline of bioethics. How, for example, should bioethics' long-term failure to address the weapons implications of genetics be explained? One possibility is that bioethicists were, for a long time anyway, simply unaware of the weapons potential of genetics. However, if this is correct, why didn't scientists and policymakers bring such issues to the attention of bioethicists working on ELSI issues? If bioethicists (who are not usually scientists) were not aware of the reality and seriousness of the weapons potential of genetics, this is presumably at least partly because no one made them aware. Part of the explanation why such dangers were not highlighted earlier may be that the spectre of biological weapons was largely overshadowed (in the minds of scientists, other academics, and policymakers) by the nuclear threat during the Cold War.[8] Be that as it may, this would not imply that the biological weapons threat should have been considered less far-fetched and worthy of discussion than many of the (often largely science-fiction) issues that bioethicists have focused on[9] and which one might also have expected to be overshadowed by more pressing concerns. Another reason may be that most scientists have themselves been largely unaware of the dual-use phenomenon. Such lack of awareness has been demonstrated by empirical research.[10] (One would have thought, however, that enough scientists would have been conscious of the weapons potential of genetics to alert bioethicists to potential dangers. As indicated above, at least those involved in the Department of Energy and Los Alamos laboratories should have been aware of the possibility, so why would they not have brought the issue to the attention of those concerned with ethical issues associated with genetics?)

A disturbing possibility is that a conscious decision was made by leading scientists not to raise the issue of biological weapons at the Asilomar conference during the 1970s[11] and that a conspiratorial silence on the part of scientists remained long afterwards. If this really is an important part of the explanation

8 Malcolm Dando, personal communication.

9 Such as much of the recent literature regarding human enhancement (for example, radical life extension).

10 Dando, M. R. and Rappert, B. 2005, 'Codes of conduct for the life sciences: Some insights from UK academia', Briefing Paper No. 16, Department of Peace Studies, University of Bradford, available: http://www.brad.ac.uk/acad/sbtwc/briefing/BP_16_2ndseries.pdf [viewed 18 October 2009].

11 Rogers, M. 1975, 'The Pandora's box congress', *Rolling Stone* (19 June), p. 37, cited in Garfinkel, M. S., Endy, D. and Epstein, G. L. 2007, *Synthetic genomics: Options for governance*, available: http://www.jcvi.org/cms/fileadmin/site/research/projects/synthetic-genomics-report/synthetic-genomics-report.pdf [viewed 5 April 2010].

of bioethics' neglect of dual-use issues, we should perhaps be wary about calls for voluntary self-governance by scientists. That is, we might be reluctant to trust scientists if they previously failed to disclose potential dangers of their work and, as a result, an important public debate about a crucial ELSI genetics topic was delayed by a decade or more.

Public debate about ELSI implications of genetics was in full force during the 1990s largely as a result of attention raised by bioethicists. However, public debate regarding dual-use implications of the life sciences did not gain prominence until early in the twenty-first century. Furthermore, this was arguably primarily due to the events of 11 September 2001 and the anthrax attacks that followed. For the most part, bioethicists neither played a major role in bringing to prominence, nor contributing to, the debates that have since ensued.

Thus, a remaining puzzle about bioethicists' lack of attention to the dual-use dilemma is why they have not further engaged in discussion of weapons implications of science since the time they most probably became aware of them. For example, the controversial experiments mentioned at the start of this chapter received a great deal of media attention, so bioethicists presumably would have heard about them. An important part of the explanation of neglect may thus be bioethicists' lack of familiarity and engagement with security issues — as opposed to clinical or medical matters — in general. The significance of the dual-use problem provides one reason why bioethicists should, in the future, become more engaged with issues pertaining to security. The security implications of infectious diseases in general provide another. Those concerned about dual-use research often advocate increased education of scientists regarding the dual-use potential of their work. However, in the aim to achieve a more informed ethical debate about research we should perhaps also advocate increased education of bioethicists regarding security. Security raises crucial bioethical issues, but bioethicists have devoted alarmingly little attention to such matters.

Ethics Discourse to Date

Despite the long lamentation above, the good news is that there have recently been at least a few exceptions to the rule that ethics literature has neglected the problem of dual-use research. There is now an emerging, growing body of explicitly ethical literature on this topic, and this is hopefully in the process of reaching a critical mass.[12] Much of the relevant literature has focused on

12 See Resnik, D. and Shamoo, A. E. 2005, 'Bioterrorism and the responsible conduct of biomedical research', *Drug Development Research*, vol. 63, pp.121–33; Green, S. K. *et al.* 2006, 'Guidelines to prevent malevolent use of biomedical research', *Cambridge Quarterly of Healthcare Ethics*, vol. 15, pp. 432–47; Selgelid, M. J.

questions about the ethical responsibilities of scientists in particular. For example, to what extent would scientists be responsible for adverse outcomes that might result from the malevolent use of their research by other actors and to what degree are they obligated to prevent misuse — perhaps by refraining from engaging in potentially dangerous research or publication when potentially dangerous discoveries are made?

Such discussion of social responsibility is important, especially given the history of scientific culture.[13] At various times in history, to a greater or lesser degree, science has been characterised as neutral, apolitical, and/or values-free. Common ideas among scientists (and others) have been that science involves an impartial pursuit of knowledge and/or that scientific knowledge is inherently good.[14] Another frequently heard idea, especially in debates about the social responsibility of scientists in the context of nuclear weapons, is that knowledge, technology and other fruits of science are neither good nor bad — but, to the contrary, it is the uses to which they are applied that are good or bad. Last but not least, it was argued that although the prevention of harmful uses of knowledge and technology may be important, scientists themselves do not have the responsibility, expertise or power to prevent malevolent applications of their work from occurring.[15] Rather than an obligation of scientists, the argument goes, the obligation to prevent harmful applications of knowledge falls on policymakers (who have — or, at least, should have — the requisite responsibility, expertise and power). If scientists do not produce anything that is inherently bad, and these other ideas are correct, one might think that scientists engaged in legitimate research are not responsible for harmful outcomes resulting from their morally neutral pursuits and products. Those who employ knowledge in a malign manner, and policymakers who fail to prevent them from doing so, would be responsible for bad outcomes; and scientists would remain innocent.

2007, 'A tale of two studies: Ethics, bioterrorism, and the censorship of science', *HastingsCenter Report*, vol. 37(3), pp. 35–43; Jones, N. 2007, 'A code of ethics for the life sciences', *Science and Engineering Ethics*, vol. 4, pp. 25–43; Miller, S. and Selgelid, M. J. 2008, *Ethical and philosophical consideration of the dual-use dilemma in the biological sciences*, Dordrecht, NE: Springer; Ehni, H. J. 2008, 'Dual use and the ethical responsibility of scientists', *Archivum Immunologiae et Therapiae Experimentalis*, vol. 56, pp.147–52; Kuhlau, F., Eriksson, S., Evers, K. and Hoglund, A. T. 2008, 'Taking due care: Moral obligations in dual-use research', *Bioethics*, vol. 22(9), pp. 477–87; Dando, M. 2009, 'Bioethicists enter the dual-use debate', *Bulletin of the Atomic Scientists*, 26 April, available: http://www.thebulletin.org/web-edition/columnists/malcolm-dando/bioethicists-enter-the-dual-use-debate [viewed 20 January 2010]; Kuhlau, F., Hoglund, A. T., Evers, K. and Eriksson, S. 2009, 'A precautionary principle for dual-use research in the life sciences', *Bioethics* (online prior to printing: doi:10.1111/j.1467-8519.2009.01740.x); Special Issue Section on The Advancement of Science and the Dilemma of Dual Use (2010), *Science and Engineering Ethics*, vol. 16(1).

13 Jones 2007, op. cit.

14 Kitcher, P. 2001, *Science, truth, and democracy*, New York: Oxford University Press.

15 Bridgeman, P. W. 1947, 'Scientists and social responsibility', *The Scientific Monthly*, vol. 65(2), August, pp. 148–54.

However, the idea that scientists should be fully divorced from responsibility for consequences of their well-intentioned research is not that tenable. If one foresees that his work is likely to be used in ways that cause more harm than good and proceeds regardless (without doing anything to forestall the harm in question), then he will be implicated in the bad consequences that ensue. If I knowingly enable a malevolent actor to cause harm, I am at least partly responsible for harm that results. We should go farther by saying that scientists have a duty to consider the uses to which their work will be applied, and that they bear significant responsibility for harmful outcomes that are foreseeable whether or not they are foreseen by the scientists in question. The point here is that scientists have a responsibility to be aware and reflect on the ways in which their work will be used. The failure to reflect or foresee the foreseeable should be considered negligence. In the context of weapons of mass destruction, such negligence could cause grave harm. If a scientist carelessly conducts and publishes dangerous research in an environment where adequate policies to prevent misuse are not in place, and a malevolent actor uses this research to cause great harm, it would be reasonable to conclude that the scientist, relevant policymakers, and (of course) the malevolent actor, are all partly responsible for damage done.

A virtue of much of the emerging dual-use ethics literature is that it takes seriously the idea that individual scientists have significant responsibilities regarding the prevention of harm resulting from malevolent use of their research. However, it might be argued that at least some authors have not taken a sufficiently clear and/or strong stand on such issues. For example, while questioning the specific obligations of scientists in the context of dual-use research, Kuhlau *et al*. (in the first paper on dual use in the journal *Bioethics*, published in 2008) argue that scientists have a duty to 'consider negative implications of research' and 'to consider whether to refrain from publishing or sharing sensitive information when the information is of such a character that it could invite misuse'.[16] They do not, however, go further by saying how one should act on his or her deliberations. The mere duty to consider the consequences of one's actions is presumably too weak if one is not further obligated to refrain from the actions in question if certain expectations result from the consideration in question. Not only do they fail to say how a scientist is obligated to act based on consideration of the results of a potential research project or publication, they do not clearly hold that scientists have any obligation beyond the act of consideration itself. One might have expected, for example, that scientists have obligations to consider the implications of their research and publications and obligations to refrain from the research or publication when harms are (reasonably) expected

16 Kuhlau *et al*. 2008, op. cit., pp. 484–5.

to outweigh benefits, or something like that. In the context of research, at least, perhaps this is what these authors have in mind, but it is odd they do not say so more explicitly.

In the context of publication, in any case, Kuhlau *et al.* offer resistance to the idea that scientists clearly have a strong obligation to refrain from publishing dual-use discoveries with dangerous implications, but their analysis here conflates separate issues:

> [W]e need to recognise such values as publishers' freedom of press and scientists' legal right to publish. It is therefore controversial to propose an obligation inflicting too many restrictions. Restrictions on publications have several implications, for example, for scientists' need to be able to replicate results in order to conduct further research, build upon the results of others, and develop and maintain a scientific record and reputation.[17]

The problem is that this kind of concern confuses the question of what a researcher has a moral obligation to refrain from doing with the question of what a researcher should be legally prevented from doing. Whether or not censorship by government of dual-use research would ever be called for is an important question. However, one could consistently believe there are cases where scientists would have a moral obligation to refrain from publication without thinking the obligation should be enforced by law or governmental censorship. The legal right to freedom to publish is not incompatible with a moral obligation not to publish because not all obligations are (or should be) enforced by law. What one is morally required to do and what one is legally required to do are distinct but related questions that should be treated separately. I might think I clearly have a strong moral obligation to walk my dog twice a day and say my prayers before I go to bed at night, but this would not imply that I think the law should require me or anyone else to do such things. Proposing that scientists have obligations to refrain from publication in problematic cases simply does not entail support of censorship, as is suggested by Kuhlau *et al.*[18]

---

17 Ibid.

18 Kuhlau *et al.* conclude (p. 485) that 'the duty not to publish or share sensitive information' is 'potentially reasonable, although phrased too much in the negative'. Given this, and their final suggestion that censorship by government might on occasion be called for after all, it is not entirely clear what specific duty beyond the 'duty to consider whether to refrain from publication' they are arguing for. Rather than defending or establishing a stronger specific duty of scientists in the context of publication, their argument appears to be more focused on defending self-governance of scientists, at least in most cases.

The Way Forward

Two things remain to be said about the apparent focus of much of the emerging ethical literature on the duties and responsibilities of individual scientists. First, although the dual-use phenomenon undoubtedly raises crucial ethical questions about the duties and responsibilities of individual scientists, it is by no means an ethical issue for scientists alone. The phenomenon of dual-use research, science, and technology also calls for important ethical decision-making by actors (with duties and responsibilities) at other levels. Research institutions (insofar as they are at liberty) must decide how to oversee activities within their confines and whether or not to provide (and perhaps require) relevant education. Scientific associations need to decide whether or not and/or how to address dual-use research in codes of conduct; and they must decide whether or not and/or how to enforce such codes on members. Publishers need to decide what to publish and/or what screening mechanisms to put into place. And governments must decide whether or not and/or how to impose restrictions on dual-use research and technology. Governmental regulations could, among other things, potentially call for mandatory reporting of dual-use research to committees for clearance before experiments are conducted or published and/or compulsory education of researchers about the dual-use phenomenon and/or ethics. Finally, funders of scientific research must decide what research to fund; and they must decide whether or not relevant education, adherence to codes of conduct and/or reporting of dual-use research to committees before experiments are conducted or published should be conditions of individual researchers' or research institutions' eligibility for funding. The dual-use phenomenon raises ethical issues for decision-makers at each of these levels, because they all face the ethical question about how to strike a balance between the protection of security and the promotion of academic freedom and/or scientific progress (assuming these things will sometimes come into conflict[19]). More detailed ethical analysis of the responsibilities of these other actors is therefore important.

Although governmental regulation of research is controversial for reasons considered at the beginning of this paper, it may be imprudent to rely too heavily on voluntary governance of scientists or the scientific community — even if we gain more clarity about the social responsibilities of scientists. One reason that mandatory measures might be called for is that scientists may not always have sufficient expertise for judging the security dangers that might result from their research and/or publications. Responsible decision-making requires assessment of the security risks and social benefits likely to arise from any given experiment or publication. Scientists, however, usually lack training in security studies and thus have no special expertise for assessing security risks

19 Some might argue that free or open science would provide the best means to maximisation of security.

in particular. In some cases they are systematically denied access to information crucial to risk assessment. For example, in the case of the mousepox study a primary concern was the possibility of proliferation of smallpox from former Soviet weapons stockpiles of the virus — that is, because bioweaponeers would need access to the virus in order to apply the mousepox genetic engineering technique to it in the hope of producing a vaccine-resistant strain of smallpox. However, any detailed information about smallpox proliferation is classified information to which the vast majority of scientists would not have access. Thus, in this important case, which has been a paradigm example of dual-use research of concern, ordinary scientists would be unable to make an informed assessment of the risks of publication.

An additional reason not to rely too heavily on voluntary self-governance is that conflicts of interest may often come into play. For example, given that career advancement in science is largely determined by publication record, a researcher may often have self-interested reasons for publishing potentially dangerous findings even when this might not be in society's best interests, all things considered.[20]

A second reason why ethical analysis of dual-use research should not focus too heavily on social responsibilities of scientists is that their duties (regarding whether or not to pursue a particular path of research or publication) cannot be determined in a vacuum. What exactly an individual should or should not do partly depends on actions taken by other actors at other levels in the science governance hierarchy.

Given the ultimate aim to avoid the malevolent use of dual-use technologies, it is important to recognise various stages in the 'dual-use pipeline' where preventative activities might take place, or regulations might operate. First, there is the conduct of research that leads to dual-use discoveries. One way to prevent malevolent use is thus to prevent the most worrisome experiments from taking place to begin with. A second way to prevent malevolent use would be to prevent dissemination of dangerous discoveries after they are made — that is, by not publishing them oneself (self-censorship), or by stopping others from publishing them (censorship). A third way would be to prevent malevolent use by limiting who has access to dual-use technologies and materials such as 'select agents' or potentially dangerous DNA sequences, requiring licensing of those using such technologies and materials, registration of relevant equipment, and so forth. A fourth way would be to strengthen the Biological and Toxin Weapons

20 Selgelid, M. J. 2007, 'A tale of two studies: Ethics, bioterrorism, and the censorship of science', *HastingsCenter Report*, vol. 37(3), pp. 35–43.

Convention via the addition of verification measures. This would at least help prevent state actors from using legitimate science for the promotion of offensive biological-weapons programmes.

The point here is that the question of whether or not a researcher has a duty to refrain from pursuing a particular project or publishing a particular study partly depends on what preventative mechanisms are in place further down the 'dual-use pipeline'. For example, if one discovers how to synthesise a particularly contagious and/or virulent pathogen, the propriety of publishing this partly depends on whether regulatory measures that would prevent this finding from being employed by malevolent actors have been implemented. For instance, if there were stronger controls over access to the technologies and materials (for example, DNA sequences) required by others to reproduce such a pathogen and/or if the BTWC was strengthened by the addition of verification measures, the dangers of malevolent use arising from such a publication would be lower than otherwise. Thus, whether or not a researcher would have a duty not to publish in such a scenario, assuming they were at liberty to do so, at least partly depends on whether or not policymakers have fulfilled their duties to put adequate preventative measures in place.

In addition to further expanding discussion beyond the responsibilities of individual scientists, there are additional fruits to hope for when ethicists more actively engage with the dual-use problem. There are obvious ways in which it raises issues similar to (or overlapping with) those discussed in ethical debates about the doctrine of double effect and the precautionary principle. Whether or not either of these is plausible or correct, the well-developed discourse surrounding them would presumably shed light on the ethics of dual-use research. Rational decision theory and discourse about 'acts and omissions' likewise address relevant issues. The point is that a long, rich history of ethical debate in these and other areas might fruitfully be brought to bear on the dual-use problem if those with expertise in these and other areas of ethics only applied their minds to it. To date, it is safe to say, rigorous ethical analysis of the dual-use dilemma has only scratched the surface.

Chapter 2: Educating Scientists about Biosecurity: Lessons from Medicine and Business

JUDI STURE

When looking at the intersection of ethics and biosecurity, we are generally concerned about how we may highlight ethical issues and solutions as a means of mitigating the risks of biotechnology being used for malign purposes. This chapter sets out to discover what we may learn for this endeavour from attempts to teach and develop ethical practice and awareness in the fields of medicine and business.

These two areas have paid considerable attention to the teaching and development of ethics in practice while also addressing social, professional and national cultures, which is a key factor in the recognition and interpretation of ethical issues. Because these sectors are engaged on a daily basis with two of the most accountable areas of human experience — health and money — they are probably the focus of the highest degree of ethically related litigation and risk of incurred costs around the world. No doubt this is at least part of the reason behind their increased attention to ethics in recent times. However, some situations go far beyond the issue of economics and law, at least in theory and aspiration, and grow out of global concerns that focus on a common, shared humanity and the goal of human safety and security. By reviewing work from these areas, the aim is to present several recommendations as to how bioethics education around issues of biosecurity in the life sciences may be most effectively addressed alongside existing ethics education courses within life-science degree programmes.

Experiences with Ethics

It is important to recognise that life scientists already work in an atmosphere of ethical awareness and accountability. However, recent research[1] in Europe shows that while ethics and biosecurity are a part of some university-based education and training in the life sciences, they are typically only a very small part, often viewed as a 'bolt-on' concept rather than an integral part of the professional identity of the individual scientist. It is against this background that we can make the claim that there is currently insufficient recognition among scientists of the potential risks of the destructive use of life-science research and we should recognise that this may be a significant barrier to addressing this problem.

From my own communications with colleagues teaching ethics and working on the development of ethics approval and monitoring processes in the UK, it appears to be commonly believed among scientists and others working in academia that society's ever-increasing concerns about ethics and responsible research have been adequately, if not too heavily, addressed by the rise in the prevalence of instruments of control or guidance in these areas. Antagonism among researchers in the UK to a perceived increase in ethical-approval processes appears widespread in my experience. Typically, however, this antipathy is generally hidden from public expression. Yet, it is my estimation that there remains substantial resistance to the requirements of ethics accommodation and approval at the grassroots level, and this is probably due, amongst other things, to a failure of those driving the policies to adequately engage with professionals in the various disciplines to explain and situate the issues effectively. In the past two decades codes of practice, ethical policies and standards of ethics have proliferated in professional associations, universities, research laboratories, and in the public and private commercial sectors. But I would argue that these instruments are insufficient to address the growing risks of dual use in the global security arena. The existence of codes of ethics does not preclude the need for effective enhanced education in ethics on dual use and other biosecurity risks, as evidence shows that their existence does not equate to full or even partial compliance with them.[2]

---

1 Mancini, G. and Revill, J. 2008, *Fostering the biosecurity norm: Biosecurity education for the next generation of life scientists*, report by the Landau Network-Centro Volta and Bradford Disarmament Research Centre; Dando, M. and Revill, J. 2010, 'Building international educational resources', this volume.

2 Kaptein, M. and Wempe, J. 1998, 'Twelve gordian knots when developing an organizational code of ethics', *Journal of Business Ethics*, vol. 17(8), pp. 853–869; Schwartz, M. 2001, 'The nature of the relationship between corporate codes of ethics and behaviour', *Journal of Business Ethics*, vol. 32(3), pp. 247–62; Schwartz, M. 2002. 'A code of ethics for corporate code of ethics', *Journal of Business Ethics*, vol. 41(1–2), pp. 27-43; Schwartz, M. 2004. 'Effective Corporate codes of ethics: Perceptions of code users', *Journal of Business Ethics*, vol. 55(4), pp. 321–41.

Most educated people with whom I work and interact in the UK and beyond (including researchers, tutors and students from around the world) operate on the assumption that they already 'know what ethics are' and know how to 'act ethically' in their work. However, when asked to name a few ethical principles, to describe and locate themselves within the framework of any theory of ethics in research, most fall silent. What people really mean is that they know they already have a value-set. The confusion arises because they are typically referring to their private value-set rather than their professional ethical standards. Unsurprisingly, people become defensive when they feel challenged about their ethical principles, but they mistake the challenge for an assault on, or a questioning of, their private values. Resentment against the teaching of ethics to students and professionals is largely a result of the failure of ethicists to explain clearly what it is they are really referring to. Ethics seems to be one of those unusual subject areas in which most people believe they do not need to be educated because they already understand 'it'. Clearly, this is not always the case. I would argue that the current burgeoning of codes and guidelines can actually blind people into thinking that ethical awareness can be reduced to a tick-box activity rather than being an element of professional identity, character and responsibility. It is too often viewed as an added extra, or something to consider in case of audit.

I would make a case that as well as limited engagement with professional ethics in the context of biosecurity within the life sciences, there is insufficient recognition among scientists in general of the role of private ethical values in the personal uptake (the 'buy-in') of professional ethical ideals and views.[3] We are all shaped by the culturally derived and expressed standards that governed our upbringing but we seem to forget that these varying value-sets may not always align with professionally required or assumed principles. Einstein apparently said that common sense is the collection of prejudices we have acquired by the age of 18, but this tends to be overlooked as we pat ourselves on the back for our educated and objective outlook as scientists. There is arguably too much emphasis and reliance on the notion that scientists are a breed apart from all cultures, shaped by and sharing in a set of values that are neutral, truth-pursuing and non-judgemental, as if being a scientist somehow confers on a consenting individual a new set of cultural values that supersede all those previously held. This is patently nonsensical, and cannot possibly be true unless the education of scientists somehow produces humans who have never been, and are no longer, subject to the forces of human nature, along with prevailing and past cultural pressures.

3 Sture, J. 2010, 'Private morals and public ethics: cultural aspects of ethical development and ethical learning in the scientific context', Paper presented at the 'Promoting Dual-Use Ethics' Workshop, Australian National University, Canberra, 28–29 January 2010.

Nevertheless, it is surprising how such a view persists and appears to be held dear by scientists themselves around the world in relation to their day-to-day work. In practice, this view is often a confusion between impartiality and neutrality[4] or, at the very least, an attempt to secure value-free 'safe' knowledge that does not 'tread upon the sensitive ground of politics or ethics'.[5] To question this sacred truth is seen to challenge the very nature of the scientific endeavour, rather akin to any attempt to question the right to free speech or academic freedom. However, concerns about the destructive use of the life sciences need to be addressed. If this involves challenging some long-held but perhaps mistaken notions, then so be it.

The emergent appreciation of the risks of biosecurity and biotechnology research was highlighted by the United Nations in December 2008 when the Meeting of States Parties to the Biological and Toxin Weapons Convention (BTWC) called for investigation into and development of the education of life scientists and relevant stakeholders.[6] If concern is being raised at this level, and is supported by a growing body of evidence suggesting that life sciences and the recognition of dual-use risks is not sufficient among practicing scientists, it seems reasonable to consider how we may go about responding. While the antagonism among many professionals to further development of codes of ethics and approval processes remains, I would suggest that we go beyond the route of codes of ethics and practice (which appear to be the commonly followed path of ethics education and monitoring) and further develop existing ethics education frameworks as a means to communicate the potential for ethical approaches to assist professionals in addressing a range of biosecurity challenges. The next section reviews how this might be done by considering lessons from arenas that have studied matters of ethics for some time.

Ethics Education in Medicine and Business

The most commonly recognised areas in which ethics play a huge part in regulating the behaviour of practitioners are those of medicine and business. My review here of a range of papers from the journal *Academic Medicine* focuses on ethics education of students in US medical schools in the 1990s and early 2000s. My appraisal of work in the business and management literature focuses on the *Journal of Business Ethics*. In contrast to the vast amount of literature

---

4 Lacey, H. 2004, *Is science value-free? Values and scientific understanding*, London: Routledge.

5 Proctor, R. 1991, *Value-free science? Purity and power in modern knowledge*, Cambridge: Harvard University Press.

6 United Nations 2008, *Report of the meeting of states parties*, December 2008, Convention on the Prohibition of the Development, Production and Stockpiling of Bacteriological (Biological) and Toxin Weapons and on their Destruction, Geneva: UN Publications.

on medical ethics and the training of doctors as distinct subject areas, ethics in the business world tends to be addressed under umbrella concepts such as Corporate Social Responsibility (CSR), although it is commonly recognised as a discrete issue in companies seeking to synchronise ethical standards in widely differing cultural settings. Crucially, the business concept of CSR often transfers ethical responsibility onto the company rather than the individual, providing a contrast with the medical approach that focuses on the individual practitioner. This is useful when considering the possible destructive uses of life-science research as we need to consider not only the role of the individual but also the community or group, whether it be a professional association, some cultural entity or a nation state.

By looking at the education and development of professionals in these contexts, it is possible to find some useful themes that may help us as we move towards enhancing life scientists' ethics skills in settings where otherwise beneficial research may be subverted for malign purposes, by equipping them to build a sustainable capacity in understanding and pass on the baton to subsequent generations. In the following sections I highlight points drawn from these areas that may be of use in supporting the development of ethics education among life scientists as a means to enhance our security.

Theme One: The Hidden and Informal Curriculum

Ethics education — at least at some level — is already part of many science education programmes, even if it largely fails to address the risks of malign use of research. It is expected that students and professionals hold ethically appropriate and responsible views, but work has shown that competing value-sets and pressures to which they are exposed during education can undermine individuals' ethical standards. What is explicitly taught is not always what is learned.

A key finding among studies in US medical schools was the existence of a 'hidden curriculum' that pervaded education, often to the detriment of ethical behaviour among students and junior doctors. A number of researchers[7] found that despite the stated commitments of the medical education system to patient well-being, altruism, empathy and caring, another value-set was being promoted tacitly. This promoted detachment, self-interest, objectivity and a business outlook among the students and newly qualified doctors. The two value-sets were in conflict.

7 Hafferty, F. W. and Franks, R. F. 1994, 'The hidden curriculum, ethics teaching, and the structure of medical education', *Academic Medicine*, vol. 69(11), pp. 861–71; Hundert, E. M., Hafferty, F. W. and Christakis, D. 1996, 'Characteristics of the informal curriculum and trainees' Ethical choices', *Academic Medicine*, vol. 71(6), pp. 624–33; Hafferty, F. W. 1998, 'Beyond curriculum reform: confronting medicine's hidden curriculum', *Academic Medicine*, vol. 73(4), pp. 403–7; Coulehan, J. and Williams, P. C. 2001, 'Vanquishing virtue: The impact of medical education', *Academic Medicine*, vol. 76(6), pp. 598–604.

Hafferty[8] concluded that by looking at the 'products' of the medical education system — be they courses, buildings, appointments of faculty, and so on — it was possible to see evidence of the implicit, business-led culture dominating the medical education system through financial values rather than patient-oriented focus. This even extended onto wards, with senior doctors and medical teams requiring students to behave in certain ways in order to 'get on' professionally, concentrating on the needs of the student doctor instead of the patient. Not only did students absorb these implicit values from faculty and senior doctors, but also they passed them on between themselves. Resultant changes in student behaviour comprised an erosion of the ethical 'being a doctor' standards that they had held when they entered medical school. Students reported having to move away from the traditional empathetic and patient-oriented perspective towards a career-building, self-advancing, financially motivated strategy.

In relation to the overall concerns of this volume, this may lead us to consider what messages about ethics in research are being sent out and heard by life-science students in the laboratory and classroom. When we teach ethics in life sciences we need to beware of tacitly compromising this with career-driving values. It is understandable that financial pressures affect aspects of education but, while recognising this happens, we need to balance the effects by addressing the resulting ethical implications. We also need to consider how we prioritise values in practice as well as in theory, honestly admitting the pressures we are requiring students to respond to in order to progress their careers. The findings of Hafferty[9] about 'products' of the educational system are applicable to the science education system also. It is not just in the classroom that values are learned, and the 'wrong' standards may be passed on unwittingly.

While US medical schools focused on teaching and developing ethical awareness in a monocultural way (to 'fit' western standards and values), in 1997 Vega[10] questioned the tendency of US business schools to teach a universal set of ethical standards, claiming that to do so could have a negative effect on the practices of graduates. Her interest was in promoting intercultural business outcomes and she believed that a focus on westernised approaches was self-defeating in the international business world. She suggested that a combination of relevant stakeholder input, deontology, and utilitarianism could be combined with pertinent community norms, and that when applied in practice, the amount of relativism involved in making decisions could be reduced. In the ethical context, this would mean that decisions made would be situationally, culturally and ethically contingent. She proposed an approach of 'common-norming', in

8 Hafferty 1998, op cit.

9 Ibid.

10 Vega, G. 1997, 'Caveat emptor: Ethical chauvinism in the global economy', *Journal of Business Ethics*, vol. 16, pp. 1353–63.

which intercultural co-operation, shared designing of programmes, co-working in difficult teams, and a rising above 'parochialism' would allow different value systems to 'provide the continuum for bridging cultural and ethical differences and birthing mutually acceptable hypernorms'.[11] Hypernorms were defined as 'fundamental principles of ethical behaviour that guide religious, philosophical and cultural beliefs'.[12] Common-norming was defined as the moderating of one set of ethical and cultural values to meet another value-set at a mutually acceptable midpoint.

Weaver also critiqued US corporate ethical practices.[13] He emphasised the need for recognition of cultural and organisational values and traditions in 'other' cultural settings, and showed how ethical processes can be compromised and de-legitimised by culturally careless practices. This would, of course, be a potentially disastrous situation if it were reproduced in current attempts to address dual-use and biosecurity issues. Weaver's work highlights the need not only to recognise intercultural issues but also the practical and hidden 'workings' of organisations themselves in order to achieve shared understanding.

Work by Muijen[14] and Wines[15] showed that two commonly taught strategies for business-ethics training — those of compliance and a cultural change — required top-down transformation to assure their uptake by academics and students. Muijen proposed a 'third route' towards sharing cultural narratives through dialogue, focusing on empowerment and the integration (not management) of diversity of values and perspectives. She challenged the notion of CSR and questioned its meaning to differing cultures. Wines concluded that teaching students utilitarian (outcome-based) approaches is insufficient to prepare them for the complex choices that will face them in the real world. He proposed an integration of ethics with other concepts and theories to enhance students' understanding of the social and cultural place of business and ethics. These include ethical psychology, organisational design and behaviour, motivational theory, and courses on how business, society and the law interact, plus socio-political theory and the construction of regulatory frameworks. In rapidly advancing settings such as biotechnology, with its potential for harm as well as good, these are important points. What seems obvious to western eyes in concepts such as CSR is not always so clear to other cultures that may take a

---

11 Ibid, p.1361.

12 Ibid, p.1353.

13 Weaver, G. 2001, 'Ethics programs in global businesses: Culture's role in managing ethics', *Journal of Business Ethics*, vol. 30, pp. 3–16.

14 Muijen, H. 2004, 'Corporate social responsibility starts at university', *Journal of Business Ethics*, vol. 53, pp. 235–46.

15 Wines, W. 2008, 'Seven pillars of business ethics: Toward a comprehensive framework', *Journal of Business Ethics*, vol. 79, pp. 483–99.

different approach to business ethics with cultural responsibility. The notion of integrating ethics education with a range of associated subjects echoes the calls made for the promotion of socially holistic ethics in medical schools.

Theme Two: Personal Morals, Professional Ethical Standards and Power Relationships

Scientists come to their professional education and practice with a fully formed value-set derived from their own cultural background. There is evidence that this pre-existing set of ethical standards may conflict with the professional principles to which they are expected to adhere. The personal moral value-set includes drivers such as social and religious attitudes and beliefs, and these may not accord with the 'value-free' drivers of scientific activities. On the other hand, certain private moral perspectives may enhance the uptake of professional ethical standards.

An example of this is found in work by Vitell and colleagues.[16] These authors looked at the relationship between religiosity and ethical identity in the individual (they define religiosity as 'the degree to which an individual is a religious person apart from his/her particular religious beliefs and the way that those beliefs are manifested').[17] They suggested these factors may impact on ethical decision-making in business and act as antecedents to the process. Whether or not we accept that religions are the source of morality, we can recognise that they prescribe principles and moral codes that can be fundamental in guiding the life of the individual. Those who hold the view that religious belief should have no role in the activities of the scientist will critique much of this. However, one does not need to be religious to recognise the power of shame and guilt as conformity-drivers. This can work to the good and the not so good. The ethics of compliance depend very much on what is being followed. Perhaps the most useful finding from this work is the recognition that religiosity can be a useful tool in understanding the motivations of an individual or group as an antecedent to ethical decision-making. One need not share the religious beliefs of people to appreciate how their views shape their perspective and may influence their behaviour.

In addition to the potential conflicts between personal morals and professional ethics, there is often confusion among professionals as to the nature of 'ethics' in practice. Is it a toolkit with which to tackle difficult challenges in the real world, or something more — perhaps a part of the professional character, or an essential part of professional identity?

---

16 Vitell, S., Bing, M., Davison, H. K., Ammeter, A., Garner, B. and Novicevic, M. 2009, 'Religiosity and ethical identity: The mediating role of self-control', *Journal of Business Ethics*, vol. 88, pp. 601–13.
17 Ibid, p. 602.

Many professionals arguably never face the potential inconsistencies between their personal ethical values and the professional standards they tacitly uphold because they either do not recognise the conflict or such a variance never arises in their work. I would suggest that 'ethical awareness and practice should be developed as an embedded part of being a [scientist]' (adapted from Hafferty and Franks[18]). Evidence from the medical-school context showed that students were unable to operate according to the ethical standards to which they aspired because of the pressures put on them by the system.[19] It was suggested this resulted from the relatively powerless position of students in the system hierarchy. Feudtner, Christakis and Christakis proposed that more effort should be directed at maintaining ethical standards rather than trying to alter behaviour, and this could be helped by timely, practical guidance from seniors in dealing with difficult cases as they arise. This placed responsibility on students to raise issues, and seniors to respond effectively to them — meaning that seniors needed to be ethically aware and competent too, often learning from juniors.

In the framework of the concerns of this volume, this leads us to consider how experienced tutors, researchers and practising scientists embody their ethical standards in relation to their private views on a daily basis. Are we practising what we preach? Do we facilitate the provision of adequate and appropriate safe space and time in which students, tutors and working scientists can raise, discuss and question ethical situations and dilemmas without sanctions?

The business literature examined for this chapter also reflected on the potential conflict in power relationships when it considered the ethical expectations of parent companies and subsidiaries operating in different cultures. This work focused on the workplace rather than the educational establishment. Thorne and Bartholomew Saunders[20] showed how cultural values affected ethical reasoning in multinational companies. They concluded that businesses must not ignore cultural variations in ethical perceptions. They suggested that companies should integrate working systems in such a way as to underpin their global corporate goals while still responding to local organisational norms and routines. They proposed that ethics policy-making teams should comprise people reflecting the full cultural diversity of companies' business operations.

---

18 Hafferty and Franks 1994, op cit.

19 Feudtner, C., Christakis, D. A. and Christakis, N. A. 1994, 'Do clinical clerks suffer ethical erosion? Students' perceptions of their ethical environment and personal development', *Academic Medicine*, vol. 69(8), pp. 670–9.

20 Thorne, L. and Bartholomew Saunders, S. 2002, 'The socio-cultural embeddedness of individuals' ethical reasoning in organizations (cross-cultural ethics)', *Journal of Business Ethics*, vol. 35, pp. 1–14.

Work by Robertson and Fadil[21] and Kim and Kim[22] illustrated how cultural aspects of belief and behaviour in the business world can influence the formation of ethical standards. Both ethical intensity and an alignment of cultural with ethical values were instrumental in mediating the decision-making of individuals in the professional context. Moral intensity[23] is defined as the variance in peoples' response depending on the intensity of an ethical dilemma, meaning that more effort is usually put into the hard decisions than easier ones. This of course presupposes that the difficulty or seriousness of a situation is recognised effectively in the first instance. Kim and Kim, in their work on Korean public-relations professionals, found that Korean values of social traditionalism were significantly involved in explaining professionals' attitudes to CSR. Because traditional Korean values harmonised with much of the overall CSR conceptual framework, it was much easier to achieve a good uptake of those values in that culture. Therefore, arguably this alignment factor should be incorporated into any attempt to construct a cross-cultural professional-ethics system, particularly in a potential dual-use context. By taking advantage of cultural and professional ethical alignments in particular frameworks, it may be easier to gain a widespread common agreement about a specific ethical situation.

As with work in US medical schools that commented on the lowly status of medical students, research by Secchi[24] recognised that individuals in companies are typically identified by their status within the hierarchy and the tasks they have to undertake. Secchi also considered other personal characteristics including culture, gender, age and attitudes towards politics, the environment, religion, and so on. His study resulted in the identification of four implications for business ethics and social responsibility in practice that may be translated into other arenas.

He found that everyone is 'ethically aware' as a means to self-advancement and this is enhanced when people regularly engage in the same social channels (workplace). Individuals best develop a sense of the repercussions of their actions when they interact frequently in situations that enable them to build a cognitive picture of the positive outcomes that can benefit them. In relation to the concerns of this volume, this might involve scientists in two scenarios. Firstly, the social channels of the classroom, laboratory, the grant-writing desk and human-resources department, all of which largely dictate the scientist's current and future prospects, status and financial security. Secondly, the social

---

21 Robertson, C. and Fadil, P. 1999, 'Ethical decision making in multinational organizations: A culture-based model', *Journal of Business Ethics*, vol. 19, pp. 385–92.

22 Kim, Y. and Kim, S.-Y. 2010, 'The Influence of cultural values of perceptions of corporate social responsibility: application of Hofstede's dimensions to Korean public relations practitioners', *Journal of Business Ethics*, vol. 91, pp. 485–500.

23 Robertson and Fadil 1999, op. cit.

24 Secchi, D. 2009, 'The cognitive side of social responsibility', *Journal of Business Ethics,* vol. 88, pp. 565–81.

channel of the communal environment as a safe forum in which to discuss potential or real dilemmas and articulate and respond to these with the input and support of peers, instead of feeling isolated, unsure or afraid of the best way to respond in challenging situations. This latter point is visibly lacking in the day-to-day science context.

Further, Secchi found that social responsibility (behaving ethically) serves as a reinforcing mechanism. It works as a shared 'tie' between the giver and receiver (colleagues, end-users, employers, stakeholders) and it is up to the individual to be socially responsible or not. However, as he pointed out, there are difficulties in this scenario. The freedom to choose how to act can be impacted by a range of forces. At the bottom of this is motivation, and it is obvious that individuals can appear to act ethically when in fact they are only doing so for self-serving purposes. The apparently ethical action may not be driven by altruism, philanthropy or some other beneficent force but purely or principally by self-interest. This makes it impossible to truly interpret the action of the individual, and in this we confront the personnel reliability dilemma. However, unless we allow each scientist the freedom to choose how to act, we are exposing ourselves to the prospect of totalitarian control of the scientific process, which cannot be a viable or desirable way forward. The final finding of interest made by Secchi is the effect of the long-term exploitation of social channels experienced by individuals in groups. This presupposes the availability of shared social channels in which each person can formulate, test and enhance their sense of social responsibility or ethical perspectives in a safe and supportive space.

The life sciences are an area of research that could potentially produce significant malign outcomes for the world's population. This should lead us to consider the need for bodies such as universities and research-science organisations to explicitly recognise the existence of culturally diverse values and norms by enabling and requiring discussion and active focus on them in relation to ethics in science. However, while aiming to encompass a range of cultural perspectives, we should clarify that recognition need not, and should not, necessarily mean validation or result in practical implementation.

While this should not be difficult in view of the present-day emphasis on diversity, in practice, it will probably be challenging because such a process may appear judgmental. In particular, this could be the case when we consider religious pressures and norms, and characteristics such as religiosity. However, I would suggest this is a nettle that needs to be grasped. If a greater understanding of religious ideals and norms as opinion-formers and behaviour-drivers will enhance our understanding of how we make ethical decisions, let's pursue that understanding. There is a need to integrate systems in our universities and scientific organisations to reflect our global ethical goals while still responding to 'local' norms; that is, culturally variant norms. We should involve representatives of all cultural stakeholders in decision-making when developing policies as well as deciding what sanctions would result if policy is not put into practice. Social channels can be any concept from the family to the state level, each applying its own pressures to behave in certain ways in order to 'get on'. Crucially, we need to introduce, or develop and support, where existing opportunities are offered, a safe time and space to engage in ethical debate within scientific communities and provide spaces where individuals can formulate, test and enhance their sense of social responsibility or ethical perspectives.

Theme Three: Ethics Training at the Right Time

Medical-school research shows that it is important to focus ethics education effectively in terms of timing and content. Instead of delivering 'classes on ethics' and testing students on their theory knowledge, it is seemingly more beneficial to stage the delivery and development of ethical awareness throughout the course of a programme.

Work by Christakis and Feudtner[25] found that traditional medical-school ethics teaching was limited in its effects because it did not focus sufficiently on the needs of students in ways that reflected their stage of training: students were being taught about their ethical responsibilities as doctors but they were still students. They concluded that ethics education for students should be 'resituated within a framework of the student's ethical development'[26] and recognition must be made of the stages of intellectual and emotional development that students pass through during their training and professional life. They also recognised that there should be a place for such changing judgements to be aired and allowed within the hierarchy of the medical team without negative repercussions. It was apparent from their work that the ethical theory or ethical principles approach needed to be augmented with daily decision-making processes and practice, to allow the theoretical to become something like a professional code of conduct. They suggested that a process-oriented model of ethics teaching would allow students to develop their ethical reasoning capabilities both individually and as part of a team.

This study was supported in 1994[27] in an examination of the effects of a single ethics class on a group of first-year medical students. It was concluded that a short course in ethics was unlikely to change students' values or opinions,

---

25 Christakis, D. A. and Feudtner, C. 1993, 'Ethics in a short white coat: the ethical dilemmas that medical students confront', *Academic Medicine* , vol. 4, pp. 249–54.

26 Ibid, p. 253.

27 Shorr, A. F., Hayes, R. P. and Finnerty, J. F. 1994, 'The effect of a class in medical ethics on first-year medical students', *Academic Medicine*, vol. 69(12), pp. 998–1000.

citing the varied backgrounds, religious affiliations and personal experiences of classes. This was echoed in the business literature by Wines,[28] who suggested that there should be a progression of ethics sessions or courses throughout an educational programme, tailored to match the level of ethical decisions that were being faced by students at each time, and topped off with a 'capstone' ethics course to complete the circle.

When considering biosecurity and the protection of life-science research from harmful uses, this may lead us to consider the timing of ethics classes as well as the content. Asking too much too soon of students may be inappropriate. By tailoring ethics teaching to a stage-appropriate set of questions and challenges, students may engage more effectively with both practical and theoretical issues. It seems that it is also as important to provide students with opportunities to develop associated interpersonal skills, as it is to provide them with ethics as a set of knowledge. By providing a longitudinal, stage-specific, culturally and philosophically holistic set of ethics courses and associated sessions and discussion opportunities, we may be able to develop in individuals and groups a greater sense of becoming a ethical practitioner and promoter of social benefit. In addition, it is important to pass on the notion that ethics is not just a set of facts, rules and principles, but a way of being and a part of identity as a professional.

Theme Four: Culturally Holistic Training

It is has been recognised in both the medical and business literature that it is not only desirable but necessary to incorporate an increased social awareness into educational programmes and professional practice. This is an issue that goes right to the heart of the risks of biotechnology being used for malign purposes. It can be argued that scientists working in research laboratories are to some extent cushioned from exposure to the outside world and the effects of their work. Today, while this may be acceptable as long as effects of the work are beneficial, it is clear that we should highlight to professionals working in life sciences the need to review their work and its potential outcomes in a wider, culturally holistic sense. It is probable that we can no longer simply carry out research for its own sake and publish freely, as has been largely the case to date. We must consider far more carefully how we work and communicate in the future. This entails consideration of how we view and approach scientific work from the beginning of any project to beyond its laboratory end.

In seeking to minimise risks of dual use this may lead us to consider the bigger picture — a holistic view of where our science sits in the world. Just as medicine

28 Wines 2008, op. cit.

lies in a range of cultural frameworks, so do the life sciences. It is important to challenge students, tutors and working scientists with an outsider's view of their work. Why are they doing what they do? Who will it help? Who will it hinder? Is it right to pursue what they are doing? Clearly this is already happening in some fields, for example in relation to human reproduction and stem-cell research, but we need to widen this approach to include all life sciences. We should view ethics education as enculturation into professionalism — we need to begin to develop new norms and challenge old ones that hold we may do science because we can do science. Hafferty and Franks[29] suggested that teaching and learning should operate in a reflective and responsive way as students confront ethical issues, but that ethical issues should be considered in the wider cultural environment in which the medical tradition exists. Their view was that rather than a need for more classes, ethics would be better taught by starting training in it early and continuing throughout the learning process. An 'ethical' (that is, professionally 'moral') view of life science may need to be reinstated — and we need to negotiate the pressures that define those private-into-professional morals as ethical identity.

From the business side, Wines' ideas[30] about integrating ethics teaching with that of other concepts are also important. Unless we introduce a range of culturally meaningful theories and ideas to scientists and associated professionals, the notion of ethics will continue to be viewed as a 'bolt-on' to everyday practice. By incorporating a range of subjects into the delivery of ethics education, a holistic perspective will be provided and fostered in those engaged in learning. Wines' focus on the need to enhance ethics teaching as a means to counterbalance the financial drivers promoted and prioritised in education and the idea of ethics 'capstone courses' is one that we ought to seriously consider in the biosecurity and dual-use context.

Theme Five: Ethics as Part of Professional Identity

The process of enculturation into a concept of professional identity is mentioned in one way or another by a range of authors already cited. This process was described by Swick[31] in a paper that identified nine characteristics of a doctor that equate to a state of medical professionalism and it would be possible to relate this easily to the life-science scenario. He highlighted the subordination of the physician's (scientist's) own interests to those of others, adherence to high ethical standards, response to societal needs in a social contract, the evincing of core humanistic values (empathy, caring, and so on), self- and

29 Hafferty and Franks 1994, 'op cit.

30 Wines 2008, op. cit.

31 Swick, H. M. 2000, 'Toward a normative definition of medical professionalism', *Academic Medicine*, vol. 75(6), pp. 612–16.

group-accountability as doctors (scientists), and reflective practice. These traits could be fostered in existing educational contexts by expanding the scope of ethics to make students aware of their individual role in the ethical behaviour of the wider scientific community. Kenny, Mann and MacLeod,[32] who developed further the notion of professional character by looking at the use of role models as an educational tool, also supported this. They suggested that attempts to teach students ethics in the tool-kit format led to the ethics of character being lost. In their view, the ethical nature of the agent is central to solving ethical dilemmas, rather than the simple application of a set of principles. They suggest a return to the virtue-ethics nature of medicine as 'virtuous physicians both model good behaviour and comprehend the reasons for their choices'[33] and the apprenticeship model reflects centuries of professional education.

Hafferty[34] identified themes through which ethics education and awareness could be addressed and easily incorporated into the life sciences: organisational policy development, a re-evaluation of all processes from teaching, learning and assessment through academic appointments and organisational practices, resource allocation and prioritisation, and what he referred to as institutional slang (changes in the use of everyday language that reflected the dominant mode of thinking in the medical education system, giving examples of business-speak, illustrating the changing socio-cultural influences acting on the system). These areas of concern reflected crucial earlier work by Miles *et al.*[35] which focused on the need to address four institutional areas in order to successfully embed ethics education in medical schools —support from the dean, support from administrative centres, the development of faculty approaches to ethics, and collegial support of an ethical culture. This top-down approach has been mentioned earlier, and is absolutely pivotal in achieving wide uptake of ethical processes and attitudes.

All of this leads us to consider issues around how scientists view themselves as professionals and as ethical practitioners. We need to enhance awareness of the place of science and scientists in society, encouraging them to look at themselves as engaged actors in a moving social, economic and technological drama. Society arguably demands more accountability today than in previous decades. Given that scientists hold such a key role in the balance between beneficent and maleficent outcomes for the human race, we need to broaden the way we look at our scientific work to encompass a truly holistic social perspective. Just as ethics ought to be a fundamental consideration in all research, so should awareness

32 Kenny, N. P., Mann, K. V. and MacLeod, H. 2003, 'Role modelling in Physicians' Professional Formation: Reconsidering an Essential but Untapped Educational Strategy', *Academic Medicine*, vol. 78(12), pp. 1203–09.

33 Ibid, p. 1207.

34 Hafferty 1998, op. cit.

35 Miles, S. H., Weiss Lane, L., Bickel, J., Walker, R. M. and Cassel, C. K. 1989, 'Medical ethics education: Coming of age', *Academic Medicine*, vol. 64(12), pp. 705–14.

of biosecurity risk be clarified in scientific endeavour at all stages of research. There is also an argument for a return to an individually mediated virtue ethics to counterbalance the purely deontological or teleological approaches. Miles' work[36] looking at the need for top-down support also balances a virtue-ethics approach; that is, meeting the individual scientist halfway. I would suggest that without high-level engagement with top-down management of ethics — be it at the level of laboratory director, course leader or politician or civil servant operating at state level — any efforts to institute and influence 'coal-face' ethical identity and individuals' values and norms will be diluted or simply not taken up in many areas.

Summary of Good Practice from Medicine and Business

In the previous sections I have considered research from the medical and business contexts and drawn out some ideas that we may be able to use and develop to prevent the malicious use of biotechnology. By drawing on lessons learned by other professionals, we can now consider a range of recommendations arising from these sources.

Five themes have been highlighted:

- The hidden curriculum that can compromise ethical behaviour
- The clash of personal morals, professional ethical standards and power relationships in which private values conflict with professional principles that are supposed to be value-free
- Ethics training at the right time, to allow for stage-appropriate learning and development; culturally holistic training in which ethical behaviour and values are considered and developed in the context of wider society rather than in the rarefied atmosphere of the laboratory or classroom
- Culturally holistic training, in which scientific activity is carried out in the recognised framework of wider society, with an embedded acknowledgement of the possible affects that could result
- Ethics needs to be part of professional identity, in which students and scientists may be allowed opportunities to develop an ethical character and foster a professional identity that encourages cultural responsibility.

Within these themes we can see many practical and theoretical issues that may offer direct support to the development of dual-use ethics awareness in the life sciences. These include:

---

36 Ibid.

- activities involved in common-norming: a sharing of responsibilities between varying cultures and levels of hierarchy in any given context when planning
- the fundamental recognition of cultural and organisational values and traditions themselves in 'other' cultural settings
- the recognition that certain cultural perspectives align well with ethical ones, but vary considerably
- the necessity of top-down transformation in order to assure uptake by academics, students and professionals
- the presentation of both deontological and teleological ethical theory to students while engaging them with real-life scenarios along with consideration of a virtue-ethics approach
- the idea that more effort should be directed at maintaining ethical standards as they are derived from private moral values rather than in trying to alter behaviour
- the need to provide and support a safe community space in which to allow students and practising scientists to discuss and debate ethical issues safely without fear of personal loss of advancement
- the need to address organisational/institutional areas in order to successfully institutionalise ethics education in the life sciences as it applies to dual use.

These themes and issues have been identified through a relatively short review of the literature in just two areas. Further work will doubtless highlight more useful material on which we can start to build. While it is not a simple cut-to-fit exercise in which we can lift lessons learned elsewhere wholesale into the biotechnology-security setting, we can in all probability move forward with these lessons confident that we may avoid some of the pitfalls experienced in medicine and business.

Conclusions

It is clear that culture plays a major part in the development of our views and beliefs. Even in the substantially monoculture of the US medical school, we can see clashes of cultural values at personal and professional levels; and how much more so do we see these when comparing value-sets from around the world in the business literature?

Some may resist enhancement of ethics education to incorporate dual-use issues and argue that the subject is being overcooked and that we are in danger of 'seeing reds under every bed'. That may be so, and perhaps only time will tell, once we have that great lens of hindsight through which to look back. But I would suggest that even if we do not eventually see as much evidence of risk as we think we face from today's perspective, we must recognise that we only need one or two cases of the malicious use of today's otherwise benign biotechnology to potentially cause untold damage to millions of people. When the possible fields in which such dual use can occur are outlined (synthetic biology, nanotechnology, neuroscience, phytopathology, to name but a few), it is surely not difficult to appreciate that an over-perception of risk, if it is such, can be justified, when one considers the potential outcomes if it is not mitigated in some way.

Motivation is a key issue in much of what we have covered above. It is too simplistic to categorise doctors as behaving ethically because of their care for the best interests of the patient and businesses acting from care for their profit margins. Doctors may just as easily be operating ethically to protect themselves from being sued for negligence as for the good of the patient. Likewise, businesses may be acting out of good will to their own employees and their dependants just as much as in the interests of maximising profits. Scientists are faced with similar dilemmas in the life-science setting, when considering how to handle risks. They can act for the greater good, but also have to consider their own career advancement, reputation, safety, and so on, as well as the cost-benefit equation for others. The scientific community can be quite closed against maverick thinkers and individuals who wish to plough a new furrow if doing so involves challenging accepted norms within disciplines or science as a whole. As academics, we may like to hold to our cherished 'academic freedom' but in reality we do not possess such a thing — everything is subject to standards and rules received and upheld by the majority to maintain the status quo.

What can medical and business ethics teach tell us about educating scientists regarding biosecurity? I would suggest we could learn a great deal, as is evident here. Perhaps the most important lesson is that we cannot simply apply the values of one culture across the board when we come to look at global issues. Even universal questions have a wide range of answers.

What then can be taken from this analysis for moving forward in relation to the concerns of this volume? I would propose we could make a decision to adopt a new approach, starting with that of a form of common-norming in the first instance, which will allow all of us space, time and an engaged audience with which to introduce, debate and disentangle any subject that may be or become of ethical concern in some way. Secondly we need to provide, as early as possible, safe places and opportunities to debate the issues without fear of repercussion. Alongside these developments we can move forward to work with life scientists to enhance a wider recognition and understanding of dual-use issues more generally. This process will take time and considerable effort on the part of many people and organisations, drawn from many cultural settings, but is surely worth pursuing as we aim to enhance existing ethical awareness among life scientists by introducing them to the concept of the dual-use of biotechnology.

PART 2

NATIONAL EXPERIENCES

Chapter 3: Linking Life Sciences with Disarmament in Switzerland

FRANÇOIS GARRAUX

Introduction

For the past few decades, the accelerated development of possibilities in engineering biological agents for specific purposes, as well as the possibility of using them with both peaceful and hostile intent, have posed fundamental challenges to security concepts at both national and international levels. In particular, the recent shift of emphasis in the nature of conflicts and the actors involved poses a challenge. Nowadays, not only states but also non-state actors are known to occasionally enforce their interests violently by the use of arms. The malevolent use of biological agents is not excluded, even though the number of occasions of real use by non-state actors remains limited.[1] Closely related to this shift, the sophistication of possible means for transporting and spreading biological agents poses a challenge. Although their weaponisation remains a particularly demanding task, the possibilities to offensively spread pathogens nowadays go well beyond traditional military weapons and munitions — as the appearance of letters filled with anthrax spores in the US illustrated in 2001.[2] In additional, challenges are posed by the genuine dual-use character of a vast number of products that derive from life-science discoveries in general and results in biological and medical research in particular.[3] Finally, these challenges are emphasised by the rapidly increased interconnection of our world ('globalisation'), resulting, amongst other things, in an almost-unlimited availability of products and information, but also in the facilitated and accelerated spread of diseases — SARS and the influenza viruses being prime examples. Such challenges open new fields and blur conceptual boundaries for

1 Jeanty, B. 2009, 'The biological weapons threat: The need for global prevention, preparedness, and response', Master Thesis, Swiss Federal Institute of Technology, Zurich, pp. 17–22.

2 Ibid. pp. 21–2.

3 Selgelid, M. and Weir, L. 2010, 'The mousepox experience. An interview with Ronald Jackson and Ian Ramshaw on dual-use research', *European Molecular Biology Organization EMBO reports*, vol. 11(1), p. 18.

international, national and human security. They demand new assessments of threats and their origins, and ultimately call for new or additional approaches when developing strategies to respond and ensure security. Also, the international disarmament community faces the pressure to open up from the traditional state–military focus, to address diverse aspects of civil life, and develop novel measures to minimise potential security risks.[4] When it comes to weapons of mass destruction in general and biological weapons in particular, the debates in recent years have increasingly focused on research by life scientists. Such debates have usually taken place in the context of 'dual use', a term that offers numerous definitions. In the framework of arms control, disarmament and the non-proliferation of weapons of mass destruction, 'dual use' is usually defined as the possibility that knowledge, facilities and technologies associated with civil applications may be used for the development, production, use or enhancement of military capabilities.[5] Anne-Charlotte Merrell Wetterwik's example of the use of sophisticated ventilation filters in a pharmaceutical laboratory for the production of a biological weapon illustrates this definition.[6] Awareness-raising and education, particularly on the potential misuse of originally well-intended research results and infrastructure, have thus been repeatedly mentioned as possible preventative measures.

In the context of its disarmament policy, the Swiss government has for years closely followed the debates on dual-use threats as well as related debates on contemporary security issues. It tries to apply the conclusions in a continuous national implementation of the Biological and Toxin Weapons Convention (BTWC)[7] and the development of security concepts. The national implementation of the BTWC, as Article IV of the Convention stipulates, also requests the prohibition and prevention of 'the development, production, stockpiling, acquisition, or retention of the agents, toxins, weapons, equipment and means of delivery specified in article I of the Convention, within the territory of such State, under its jurisdiction or under its control anywhere'.[8] A modern

---

4 A very concise introduction to the recent developments in disarmament and the need for new perspectives is given in Borrie, J. and Thornton, A. 2008, *The value of diversity in multilateral disarmament work*, New York and Geneva: UNIDIR.

5 Bonin, S. 2007, *International Biodefense Handbook 2007. An inventory of national and international biodefense practices and policies*, Crisis and Risk Network Series, Zurich: Center for Security Studies, p. 390; Resnik, D. 2009, 'What is "dual use" research? A response to Miller and Selgelid', *Science and Engineering Ethics*, vol. 15, pp. 3–5; Walker, J. 2003, 'Strengthening the BTWC. The role of the Biological and Toxin Weapons Convention in combating natural and deliberate disease outbreaks', *European Molecular Biology Organization EMBO reports*, vol. 4, special issue, pp. 61–5.

6 Merrell Wetterwik, A. C. 2009, 'Curbing illicit brokering in WMD-related items: solutions in the making', *Disarmament Forum*, vol. 3, p. 17.

7 The full title being 'Convention on the Prohibition of the Development, Production and Stockpiling of Bacteriological (Biological) and Toxin Weapons and on Their Destruction', as signed at London, Moscow and Washington on 10 April 1972, and entered into force on 26 March 1975, available: http://www.unog.ch/bwc [viewed 15 January 2010].

8 Ibid. Article IV.

interpretation of this prohibition does not limit itself to a ban on 'classical' biological weapons in the sense of those intended for military use. Rather, it also addresses the activities of non-state actors in grey areas, and refers to the general availability of know-how and technology from life-science research that could be misused.

Based on this understanding, in the last few years government authorities have begun to focus on the extent to which researchers in Switzerland are aware of possible dual uses and this has led to a repeated outreach to academic institutions and the industry. The following chapter is an attempt to write a *Werkstattbericht*, a report on work in progress, on the introduction of educational aspects into national implementation measures in Switzerland's disarmament policy. Taking a government perspective, the report first sketches the relevant conditions for such an initiative, focusing on the educational framework, awareness, and national nuclear, biological and chemical (NBC) protection. It then highlights the preparation and implementation of, and lessons learned from, an awareness project carried out in 2009 by Professor Malcolm Dando (University of Bradford) and Dr. Brian Rappert (University of Exeter) and accompanied by the Swiss government.[9] Provisional thoughts on possible further steps by authorities complement this part. Finally, some concluding remarks provide tentative thoughts in the wider context, encouraging the further development of a link between life scientists and practitioners in security and disarmament.

Education and NBC Protection in Switzerland

Education and Awareness at Universities and in the Private Sector

In line with Switzerland's federal structure, the Swiss system of public higher education is characterised by a complex system of shared competences between the Swiss Confederation (national level) and the cantons (sub-national level). The Swiss Confederation, namely the Federal Department for Home Affairs and the Federal Department for Economic Affairs, oversees the institutions of higher education mainly on policy and legal aspects, and shares the main responsibility for the general promotion of research. Here, the State Secretariat for Education and Research (SER) in the Federal Department of Home Affairs also focuses on national and international matters of university education, while the Swiss

9 For a detailed description of the seminar format and organisation, see Rappert, B. 2007, Biotechnology, security and the search for limits: An inquiry into research and methods, London: Palgrave; and Rappert, B. 2009, Experimental secrets: International security, codes, and the future of research, New York: University Press of America.

Science and Technology Council acts as a consulting body for policy matters on education. However, the direct influence of the Confederation on the rules and regulations of specific institutions is limited to the two Swiss Federal Institutes of Technology (Zurich and Lausanne) and to four federal research institutes.[10] In contrast, the cantons share the main responsibility for the universities, universities of applied sciences and further-education organisations. The cantons contribute substantially to the funding of the universities and have regulatory powers. Cantonal interests in university politics are coordinated through the Swiss University Conference, which simultaneously serves as the main platform linking the cantonal level and the Confederation.

This system of shared competences and different responsibilities, however, does not include a direct influence of government authorities on specific curricula; rather, it explicitly excludes it. Within the defined legal framework regulating predominantly administrative, organisational and financial matters, universities in Switzerland enjoy considerable academic autonomy and freedom of research and teaching.[11] This autonomy also applies to research establishments within or connected to universities and the institutes of technology, even if they are primarily financed through public funds.

Besides research carried out in institutions of public higher education, life-sciences research conducted by the private sector plays an important role. As the pharmaceutical and chemical industries in Switzerland are among the most important economic sectors, specific research becomes vital for private enterprises and is actively supported. The Novartis Research Foundation or Roche's research and development activities may serve as prime examples.[12] Research results originating from, as well as education within, the industry's development laboratories are primarily meant to meet the requirements of contemporary medicine and healthcare. In addition, they are subject to the economic imperatives the respective company faces in national and international markets. Therefore, a direct influence of government authorities on education and research in the private sector is practically impossible (and would be met with substantial mistrust and resistance), as long as education and research are completed within the existing legal framework.

---

10 State Secretariat for Education and Research SER 2008, *The Swiss system of higher education* (Factsheet), available: http://www.sbf.admin.ch/htm/dokumentation/publikationen/grundlagen/factsheets/FS01_Hochschulsystem_e_2008.pdf [viewed 8 October 2009]; further information can be gathered from The Swiss Education Server Educa 2010, *Universities*, available: http://www.educa.ch/dyn/152941.asp [13 January 2010].

11 The Swiss Education Server, op. cit.

12 Novartis 2010, *Corporate Research*, available: http://www.novartis.com/research/corporate/index.shtml [viewed 13 January 2010]; Roche 2010, *Research & Development*, available: http://www.roche.com/research_and_development.htm [viewed 13 January 2010].

However, the (deliberate) lack of a government influence on the subjects taught does not automatically lead to the complete absence of concerns about misuse of life sciences. Traces of such concerns among academic circles at least implicitly exist. As far as university education is concerned, it remains unclear whether any academic course broaches the topic of dual-use research and related security implications. The following discussion of the awareness-raising project carried out in 2009 highlights findings that indicate an almost complete absence of this topic in regular life-sciences curricula. Nevertheless, a short examination of various curricula reveals that there are several academic courses touching upon this issue, particularly in the field of biomedical ethics. Yet these courses appear to focus on ethical questions for future physicians, or else highlight legal aspects. At best, (bio)security implications and consequences for the daily work in laboratories seem to be placed on the sidelines.[13]

As far as practitioners in academic or industrial laboratories are concerned, Swiss laws oblige these institutions to assign a person with sufficient professional background to oversee biological safety. In contrast, details concerning an appropriate education and training in biosafety are not regulated.[14] This certainly confirms an awareness of biosafety in laboratories, but leaves specific questions on the knowledge of biosecurity open. (The ambiguous translation of 'safety' and 'security' into the official languages contributes to this apparently absent distinction. Both 'safety' and 'security' are translated into *Sicherheit* in German, *sécurité* in French, and *sicurezza* in Italian.[15]) Also, until relatively recently government support activities for biosafety officers were limited to very informal one-day seminars conducted by the Federal Office of Environment, the Swiss Expert Committee for Biological Safety and the Federal Office of Public Health. Only in 2008 did the government initiate and fund a Biosafety Curriculum for practitioners of the public and private sector, which will be discussed later.[16]

NBC/CBRN Protection in Switzerland

An examination of the national 'Swiss NBC-Protection Strategy'[17] indirectly confirms these preliminary findings. The federal structure of Switzerland is

---

13 See for example the information provided by University of Basle 2010, *Fachbereich Medizin und Gesundheitsethik* (German), available: http://medethik.unibas.ch [viewed 13 January 2010]; University of Zurich 2010, *Institute of Biomedical Ethics*, available: http://www.ethik.uzh.ch/ibme_en.html [viewed 13 January 2010].

14 Streuli, J. 2008, 'Biosafety and Biosecurity Concepts', *Statement of Switzerland at the BWC Meeting of Experts, 19 August 2008*, available: http://www.unog.ch/bwc [viewed 9 November 2009].

15 For a discussion of this problem of terminology elsewhere, see Sawaya, D. 2009, 'Biosecurity at the OECD', in Rappert, B. and Gould, C. (eds.) *Biosecurity: Origins, transformations and practices*, London: Palgrave.

16 Streuli, J. op. cit.

17 Eidgenössische Kommission für ABC-Schutz (2007) *Strategie „ABC-Schutz Schweiz"*, available: http://www.bevoelkerungsschutz.admin.ch/internet/bs/de/home/themen/abcschutz/strategie.html [viewed 14

also mirrored in the authorities' approach to assess, prevent and respond to biological threats, together with chemical, radiological and nuclear risks. A complex web of cooperation among the Federal Office of Public Health, the Federal Commission for NBC Protection, the Federal Office for Civil Protection, the National Emergency Operations Centre, the Spiez Laboratory, the Swiss Armed Forces' NBC Centre of Competence, and the Armed Forces' Coordinated Medical Service characterises the approach on the national level. Again, these institutions closely cooperate with the cantons and municipalities, which are mainly in charge of the deployment of sensors and first responders such as police, fireguards, and first-aid providers.[18] This broad variety of actors at national and sub-national levels can create (and has created) difficulties, often based on mutual misconceptions about their respective roles, unclear tasks and redundant structures.[19] The NBC-Protection Strategy addresses these challenges and serves as a common base and guideline for prevention, intervention and coordinated leadership. It highlights the strengths and weaknesses of the current system and lists recommendations for improvement. When focusing on prevention, the strategy recommends an evaluation of existing NBC security laws, an assessment of the Confederation's and the cantons' approach to risk management (based on 14 scenarios, as set out in the annex to the strategy), and the establishment of a National NBC Protection and Coordination Office and a coordination platform for cantons.[20]

This focus on authorities and first responders illustrates that the national strategy is based on a well-developed awareness of threats originating from the accidental or deliberate release of biological agents, and the resulting ramifications for international, national and human security. However, the strategy bears the characteristics of a risk-management tool among authorities in a federal state, and does not serve as a comprehensive policy paper. Only the inclusion of the Spiez Laboratory indicates a potential and indirect link to academic and industrial life-science practitioners. Therefore, 'education' in the context of national strategy has little to do with a preventative awareness-raising among students, but is understood as being training in crisis management for responsible authorities.

Preliminary Conclusion: The Missing Link

The short elaboration on the system of public higher education in Switzerland describes an environment in which the formation of future life-science practitioners enjoys considerable academic freedom and is faced with few top-

January 2010].

18 Jeanty, B. op. cit. pp. 61–2.

19 Ibid. pp. 112–5.

20 Eidgenössische Kommission für ABC-Schutz, op. cit.

down approaches from government authorities on thematic aspects. Likewise, education and research in the private sector is guided predominantly by actual needs in medicine and healthcare, as well as by economic imperatives. It is noteworthy that neither of these educational environments shows signs of an enhanced interest in security issues, despite their proximity to the topic, the possible relevance of security concerns to them, and the academic freedom and resources at hand. On the other hand, this short analysis of the Swiss national NBC strategy indicates an immense awareness of biosecurity concerns among authorities. The preventative approach, however, focuses on government authorities and first responders, and includes few wider references to other prospective or active practitioners outside.

This seems to reveal a missing link between life-science practitioners on the one hand, and security practitioners on the other. In simple terms, there is a missing link between the graduate student of biology, and the military intelligence officer or fireguard. Both sides have difficulties imagining the perspectives and concerns of the other.

The Swiss government took these findings as a call for an initiative to bridge this gap. The publication of a brochure titled *Biology for Peace — Preventing the Misuse of Life Sciences*[21] in 2008 constituted a first step towards addressing the low awareness of security concerns within academic circles. Even minimal knowledge of obligations and debates on security at international level in combination with an increased personal responsibility among life-science practitioners should reduce the possibility of a misuse of potentially dangerous biological agents. After the brochure's publication, the timely offer of an existing project addressing precisely the gap identified in Switzerland provided a good opportunity to develop a further step in this direction.

Implementing 'Awareness Raising': The 'Life-Sciences, Security, and Dual-Use Research' Project

International Impulses

The misuse of life sciences and the associated security issues are not new to the BTWC community. In September 2002, the International Committee of the Red Cross (ICRC) launched a public appeal on biotechnology, weapons and

21 The brochure is available on the website of the State Secretariat for Economic Affairs SECO: http://www.seco.admin.ch/dokumentation/publikation/00035/02291/index.html?lang=en [viewed 15 January 2010].

humanity, calling for a 'web of prevention' among academic circles to impede biotechnology being misused for hostile purposes.[22] Despite the failure of the ICRC's appeal to generate concrete results among the BTWC community, it paved the way for the later examination of academic research perspectives within a disarmament forum.[23] Similarly, the discussions on codes of conduct for life scientists were characterised by an interest in the misuse of life sciences. Yet, as Dando highlights, '[it] has so far been unclear...to what extent this interest in potential dual-use aspects of the life sciences has led to concrete measures, particularly concerning education'.[24] This is certainly true of Switzerland, as discussed above.

The most recent and probably major impulse for an active engagement with education in Switzerland was taken from the current Intersessional Process of the BTWC and the focus on biosafety and biosecurity, as well as oversight, education and awareness-raising by the States Parties, international organisations and non-governmental organisations in 2008. The substantial discussions during the experts' meeting in August 2008 provided a fruitful basis for an exchange between life scientists and the BTWC community on concrete measures. Based on its own recent experiences at a national level, the Swiss delegation observed that '[while] governments are best placed to create the framework required, the individual researchers and their scientific and professional associations also play a crucial role'.[25] This spirit of a mutual inclusion of perspectives as well as the need to encourage awareness within academic circles through active and preventative government initiatives also entered the Meeting of States Parties in December of the same year. The final document of the meeting states:

> (26) States Parties recognized the importance of ensuring that those working in the biological sciences are aware of their obligations under the Convention and relevant national legislation and guidelines, have a clear understanding of the content, purpose and foreseeable social, environmental, health and security consequences of their activities, and are encouraged to take an active role in addressing the threats posed by the potential misuse of biological agents and toxins as weapons, including for bioterrorism. States Parties noted that formal requirements for seminars, modules or courses, including possible mandatory

---

22 ICRC (2010) *Biotechnology, weapons and humanity*, available: http://www.icrc.ch/Web/Eng/siteeng0.nsf/htmlall/bwh?OpenDocument [viewed 12 January 2010].

23 Borrie and Thornton 2008, op. cit., pp. 58–60.

24 Dando, M. 2009, 'Dual-use education for life scientists?', *Disarmament Forum*, vol. 2, p. 41.

25 Streuli 2008, op. cit.

components, in relevant scientific and engineering training programmes and continuing professional education could assist in raising awareness and in implementing the Convention.[26]

This agreed language provided a useful backdrop for a follow-up to the publication of the aforementioned brochure. The Swiss government authorities directly involved in the BTWC process considered the proposal by Dando and Rappert to conduct a series of seminars on current debates on life sciences, security, and dual-use research as very timely and relevant.

The National Legislation

Regardless of the modalities of Switzerland's participation in a project, all activities of the government require the existence of a relevant legal base. Again, the federal structure of Switzerland is mirrored in its legal system, resulting in a patchwork of national and cantonal regulations. Switzerland identifies with the monist system and treaties such as the BTWC automatically become part of the domestic legal system. However, Switzerland does not have one, single, specific act implementing the BTWC's obligations. Instead, the Convention's obligations are implemented through the sum of numerous national and cantonal legal texts covering a large spectrum of aspects, which *inter alia,* also relate to specific obligations of the BTWC.[27] Consequently, the legal base for the awareness-raising project could not be derived from a single act, but was provided by various legal foundations. In addition to the BTWC and the 2008 Report of the Meeting of States Parties, the following national laws also pertain here:

- Federal Act on Combating Communicable Human Diseases (Epidemics Act) 1970.[28] The Confederation and the cantons are obliged to implement all necessary measures to prevent and combat the transmission of such diseases. The act furthermore specifies certain containment measures and names the responsible authorities. The Federal Act on Animal Epidemics 1966[29] and the Federal Act on Agriculture 1998[30] address animal and plant aspects, respectively.

---

26 Meeting of the States Parties to the BTWC 2008, *Report of the meeting of states parties (UN document BWC/MSP/2008/5)*, Geneva: United Nations, pp. 6–7.

27 For a detailed description of the relevant legal framework on the prevention and response to biological threats, see Guery, M. 2004, *Biologischer Terrorismus in Bezug auf die Schweiz (Unter besonderer Berücksichtigung rechtlicher Aspekte)*, Zürcher Beiträge zur Sicherheitspolitik Nr. 74, Zurich: Center for Security Studies, pp. 102–14. A comprehensive list of BTWC-relevant legal texts was similarly collected in the context of a study by Scott Spence/VERTIC in relation to the implementation of the BTWC in Switzerland.

28 Swiss legislation number SR 818.101.

29 Swiss legislation number SR 916.40.

30 Swiss legislation number SR 910.1.

- Federal Act on War Material (War Material Act) 1996[31]: Article 7 stipulates the comprehensive prohibition of the development, production, brokering and acquisition, and any kind of transfer of nuclear, biological or chemical weapons. Article 34 punishes offences committed wilfully or through negligence.
- Swiss Criminal Code of 1937:[32] Article 231 punishes the transmission of a dangerous communicable human disease, whether the offence is committed wilfully or through negligence. Articles 232 to 234 equally punish the transmission of epizootic diseases, pests, and the contamination of drinking water.
- Federal Act on the Control of Dual-Use Goods and of Specific Military Goods (Goods Control Act) 1996:[33] The act creates the base to implement international agreements on respective goods, and enhances international non-binding control measures through specific national control measures. It particularly introduces the obligation to license and report the research, production and storage, as well as the (international) transfer of, dual-use goods, and enables the authorities to punish offences.
- Federal Act on the Protection of the Environment (Environmental Protection Act) of 1983:[34] Preventative measures must be taken to protect and preserve people, animals and plants as well as biological communities and habitats. Articles 29a to 29h focus on handling organisms and regulate responsibilities, licensing and reporting for activities in contained areas or for experimental releases.

Taken together, the Swiss legislation related to the implementation of the BTWC gives a comprehensive legal base covering both civil and military aspects and actors. In this sense, it addresses the ambiguous nature of biological threats and occasionally focuses directly on dual-use goods. Also of particular interest is the repeated reference to acts committed through negligence, which delegates responsibility to individual researchers. In addition, the respective responsibilities and duties of the relevant authorities to prevent the spread of diseases are clearly defined (thus setting the base for the NBC-protection strategy discussed above). However, the act leaves enough room for further preventative measures by the Confederation or the cantons. Based on these findings, implementing an awareness-raising project was a logical next step from a legal perspective.

---

31 Swiss legislation number SR 514.51.

32 Swiss legislation number SR 311.0.

33 Swiss legislation number SR 946.202.

34 Swiss legislation number SR 814.01.

Preparation for the Project

Whilst the Political Secretariat of the Federal Department of Foreign Affairs (FDFA) takes the lead regarding BTWC matters, the Federal Office of Public Health (FOPH), the State Secretariat for Economic Affairs (SECO), and various offices in the Federal Department of Defence, Civil Protection and Sport (DDPS) — including the Spiez Laboratory — contribute their expertise and participate in the decision-making processes. In late 2008, these offices jointly decided to implement the seminars in the 2009 calendar year as a next step in an emerging long-term outreach towards academic institutions.

Making use of the experience accumulated by Dando and Rappert in conducting seminars in various countries, and to minimise the authorities' influence on the content and approach, in January 2009 the Swiss authorities, in conjunction with the two researchers, defined the basis for the project, as follows:

- The target audience should primarily include graduate students, faculty staff, and practitioners. The format foresees an interactive seminar of roughly an hour and a half.
- With regard to the content and educational material, the responsibility for the implementation remains fully with the two researchers who will conduct the seminars.
- The budget to cover all of the researchers' expenses directly related to the seminars, including travel expenses, was provided by the Alfred P. Sloan Foundation.
- The Swiss authorities act primarily as door-openers, establishing the contact between the researchers and potentially interested institutions. After a first contact, details in administration and teaching would be directly organised between the two researchers and the respective institution. Thus the role of the authorities would be an accompanying one rather than a supporting one.
- The authorities will provide logistical support to the two researchers in Switzerland.

Opening Doors

The Arms Control and Disarmament Policy branch of the Federal Department of Defence provided the administrative support, sending out an introductory letter to potentially interested academic and research institutions in early 2009. Referring to the current debates on dual-use research and its security implications, the letter invited expressions of interest from academics in a free seminar on the topic, organised by the two researchers, and accompanied by representatives from the federal government. The institutions were only required to organise a suitable classroom, date and time. To avoid the time-consuming process approaching the universities' hierarchies, the introductory letter was specifically targeted to the relevant institutes or university departments. Research institutions in the private sector were addressed via their official contact details. Interested academics and practitioners were then invited to deal directly with Rappert for all further administrative details.

The response was mixed. From the 12 institutions approached, seven indicated some or great interest, and six (in alphabetical order below) then agreed to conduct the seminars:

- Friedrich Miescher Institute, Basle
- Spiez Laboratory
- Swiss Society for Microbiology
- University of Basle, Biozentrum (Department Biozentrum)
- University of Geneva, Section de Biologie
- University of Zurich, Institute of Molecular Biology

The other five institutions showed no interest, or left the introductory letter unanswered.

The seminars were held in the summer of 2009, taking place in very diverse settings, as the following brief chronological abstract highlights:

- 4 June 2009: Seminar on the occasion of the Annual Conference of the Swiss Society for Microbiology. Time constraints meant that the seminar was slotted in between two other presentations and was limited to just 30 minutes. The fact that there were only two participants also hampered the debate. This early experience revealed the need to address life scientists within their own environment, as discussed later.
- 23 June 2009: Seminar at the Section de Biologie, University of Geneva. Attended by more than two dozen students, the increasingly lively discussions indicated an interest in the topic.
- 20 August 2009 (am): Seminar at the Department Biozentrum, University of Basle. A lively discussion among some 50 participants made the seminar one of the most successful of the series.
- 20 August 2009 (pm): Seminar at the Friedrich Miescher Institute, Basle. Despite there being only five participants, the researchers and the accompanying government official were confronted with very critical questions, finally indicating a crucial divide between life-science practitioners and security practitioners in the assessment of biological threats and their origins.
- 21 August 2009 (am): Seminar at the Spiez Laboratory, followed by a train-the-trainer session. The in-house seminar for an authority directly involved

in the implementation of the BTWC and the national NBC protection allowed the participants to gather provisional conclusions for possible further steps.

- 21 August 2009 (pm): Seminar for the Institute of Molecular Biology, University of Zurich. The seminar's very active debates among approximately 30 participants indicated a substantial interest in the topic. This was followed by further questions from students after the session.

The Seminar's Framework and Resulting Debates

The fundamental issue for the seminar was the question of whether life scientists should publish research results (and to whom) if there was a potential danger that the results would attract the attention of those with dubious motives for wanting such information and (theoretically) provide them with know-how to develop biological pathogens for offensive purposes. This represents the reformulation of the crucial assessment by every scientist whether the value of a publication outweighs the potential risks. The seminar was not intended, however, to provide a definitive answer to the question but to stimulate debate and perhaps challenge existing opinions.

Dando kicked off the debate by summarising the mousepox experiment, about which very few of the participants had heard. In that experiment, researchers had genetically engineered a virus for pest-control purposes, but this ultimately resulted in the creation of a lethal virus that even killed vaccinated mice. In theory at least, this opened the possibility of genetically engineering a lethal human virus, against which vaccination would be ineffective.[35] The seminar participants were asked whether, in such circumstances, they would publish the research result and how they would come to their decision. Based on the first responses, which usually supported the idea of publication, the angle was changed to highlight possible sequences in publishing the results in various scientific journals. This model of stimulating an open debate with questions was continuously applied throughout each seminar, with reference to recent developments in biotechnology as well as occasional responses by governments such as intensified biodefence programmes or tightened control over scientists.

The responses from participants were strikingly similar throughout the seminar series. A large number found themselves confronted with a new perspective on their work. In various after-class conversations, participants repeatedly admitted that they had never previously considered a potential misuse of their research. Despite this, the clear majority ultimately felt the value of a publication outweighed the potential risks, and justified their position in various ways. Some

35 Jackson, R. *et al.* 2001, 'Expression of mouse interleukin-4 by a recombinant ectromelia virus suppresses cytolytic lymphocyte responses and overcomes genetic resistance to mousepox', *Journal of Virology*, vol. 75(3), pp. 1205–10; cited in Selgelid and Weir 2010, op. cit., pp. 18–24.

referred to publication as a basic professional requirement ('publish or perish'). Others were of the opinion that, sooner or later, potentially dangerous know-how would be published in any case; if they did not publish it, someone else would. The prospect of a governmental top-down approach to regulate research publications was generally met with substantial scepticism. When asked about threats, the participants' description of potential dangers and possible actors with malign intent remained very vague most of the time. Few narrowed their answer to terrorists (namely Al-Qaeda), and only one or two mentioned states with extensive biodefence programmes. International regimes such as the BTWC or relevant national legislation were almost never referred to by (and were seemingly unknown to) the vast majority of participants. There were no further indications that contributors had reconsidered their research activities in light of national and/or human security prior to the seminars. Nevertheless, most participants repeatedly agreed that researchers in the life sciences share a moral responsibility; living this responsibility is, however, often limited by professional and economic imperatives.

Analysis, Lessons Learned, and Possible Ways Forward

Addressing Existing Needs

A provisional analysis seems to confirm a generally low awareness of the potential misuse of dual-use research within Swiss life-science research institutions. This manifests itself not only in a repeated unawareness of often-quoted research experiments among university students in particular, but also in a frequently diffuse or narrow assessment of potential dangers and actors. While the complete absence among life scientists of references to security policy or disarmament regimes is perhaps understandable, the lack of references to relevant acts in the national legislation is noteworthy. Nevertheless, the readiness of a number of institutions to implement the seminars and distribute the 'Biology for Peace' brochure, and the interest many participants showed, seems to verify that many life-science practitioners consider further thought on dual-use research, ethics, and (moral) responsibility by researchers to be important. The existence of several courses in biomedical ethics corroborates this trend. These findings lead to the conclusion that the government initiative to support debate on dual-use research and related security issues addresses existing needs. Moreover, further steps will be necessary if sustained awareness is to be generated.

However, the government will have to consider approaches that are not perceived as interference with the freedom of research or as an obstacle to economic independence. A direct top-down approach from national authorities on the curriculum of universities or enterprises would not only generate scepticism among practising life scientists, but would also stand in sharp contrast to the federal structures on which the Swiss educational and legal systems are based. Any further outreach initiatives will have to take into account existing patterns of shared competences and cooperation. For example, approaches that inform researchers on current debates on dual use and security during established courses, and convince them to introduce the topic into their daily environment, are more likely to succeed. Similarly, another promising method includes the sensitisation of first responders and security practitioners, encouraging them to address life scientists in the context of their daily professional activities and thus act as close and credible intermediaries.

Lessons Learned

Several points raised by Dando and Rappert during the train-the-trainer session in the Spiez Laboratory provide practical hints for possible further activities. Only a few institutions are likely to create new courses or lectures specifically on dual-use research and security. The authorities' initiatives are likely to generate sustainable solutions if incentives exist for the integration of the topic into current curricula. Such motivations could include elements or even ready-made packages of teaching material. In this context, the low attendance during the first seminar (and the speakers' experience elsewhere) confirms the necessity for addressing life scientists within their regular study and work environments, rather than trying to lure them to separate or peripheral events. This was borne out by the success of ensuing seminars that were held within frameworks familiar to the scientists by taking advantage of existing seminar series. If the topics are presented openly and the questions are debated freely, without imposing solutions, this should prove fruitful and encourage further reflection. Other methods could be based on role-plays, which force participants to argue from a specific perspective (researcher, industrialist, security coordinator, publisher, military, and so on) and thus encourage them to engage with other views.[36]

Further lessons learned refer to the necessity to minimise the number of 'gatekeepers' by, wherever possible, making direct contact with the most appropriate office or department. Initial contact with potentially interested institutions should be made by a suitable communicator among the various offices concerned. For example, the response of universities or enterprises to

---

36 Dando and Rappert provide these lessons learned, principally based on their personal experience accumulated during the seminars held in various countries. The points were presented and discussed during the train-the-trainer session in the Spiez Laboratory, 21 August 2009. For teaching material, see also Rappert, B. 2009, *The life science, biosecurity and dual-use research*, available: http://projects.exeter.ac.uk/codesofconduct/BiosecuritySeminar/Education/index.htm [viewed 5 March 2010].

a first contact made by, say, the Federal Department of Defence (despite its competence and close involvement in BTWC affairs) is likely to be different from that to contact established perhaps through the Federal Office of Public Health.

Possible Ways Forward

Based on these lessons, the Swiss authorities are currently exploring possibilities for a follow-up. Admittedly, resources in this area are limited, but the seminars were carried out successfully with almost no (financial) resources. Though any further steps are likely to be similarly constrained, this does not necessarily mean a hindrance to future progress. For example, the analysis above indicates there is little need for changes in the legal framework — the relevant provisions already exist. Similarly, the educational system offers various platforms, and a substantial number of life-science researchers share at least implicitly the concerns of security practitioners, as the seminars confirmed. The existing awareness among first responders and security practitioners on national and cantonal levels provides a further important base. In summary, this indicates an existing potential, and illustrates numerous points to build on.

In practical terms, possible ways forward could include the following:

- Based on the initiative of various government offices such as the Federal Office for the Environment (FOEN) or the Federal Office of Public Health (FOPH), the recently established Biosafety Curriculum serves as a tool to harmonise the biosafety standards within Switzerland. This is implemented in courses for biosafety officers for safety levels one to three.[37] Incorporating references to the debates on dual use and security into the courses would open the possibility of spreading the word directly into research facilities via the acting biosafety officers, and follow the principle of building on existing patterns of cooperation.
- Similarly, first responders could be made more aware of the issues through the existing patterns of shared competences and cooperation in the context of the national NBC-Protection Strategy. Via the Federal Commission for NBC Protection, thoughts on a conceptual extension of preventative measures with education could be introduced, while the Coordination Platform of the Cantons could serve as a gateway to cantonal authorities and their influence particularly on universities on the one hand, and first responders on the other.

37 B-Safe 2009, *The Curriculum Biosafety — An initiative of the FOEN, the FOPH, the SUVA and the FECB*, available: http://www.b-safe.ch/?mid=1379&pid=1381&lang_id=0&lang_id=1 [viewed 22 January 2010]; and Federal Office for the Environment 2010, *Biotechnology / Activities with genetically modified or pathogenic organisms in contained use*, available: http://www.bafu.admin.ch/biotechnologie/01744/index.html?lang=en [viewed 22 January 2010].

- The militia system of the Swiss armed forces provides a further possibility. All armed forces personnel enjoy a basic education in NBC protection, particularly specialist troops such as the NBC troops or the medical corps, which are manned by civil experts working in comparable professional fields. The militia could be encouraged to actively introduce the topic into the civil workplace.
- Finally, education could be included as an element in a comprehensive political strategy on the national implementation of the BTWC. Such a strategy could serve as a guideline to the activities of authorities directly included in the BTWC process, and thus serve as a complementary element to the existing NBC-Protection Strategy (which focuses on operational aspects in the case of a crisis and includes no disarmament features, as discussed above).

Concluding Remarks: Linking Life Sciences with Disarmament

Serving as a report on work in progress, this chapter has described the implementation of an awareness-raising project on dual-use research in Switzerland as part of a continuous implementation of the BTWC. It has highlighted the lessons learned and possible ways forward. Switzerland still needs to define concrete lines of a long-term perspective. In addition, the importance of accounting for national particularities has been highlighted — the deliberately liberal environment limits government influence, and the federal structures generate an enormous number of actors and authorities. This often results in balanced and democratic solutions, but often requires time-consuming processes, and the national implementation of the BTWC is no exception.

Despite the current lack of a universally applicable implementation model, this chapter has also shown that — as is probably the case elsewhere, too — many of the prerequisites are already in place, making it easier to address some general concerns among life scientists. The fact that the final impulse was provided by a 'classical' disarmament regime did not hamper addressing actors and perspectives beyond the 'classical' disarmament horizon. Referring to recent successes in disarmament negotiations, John Borrie and Ashley Thornton extensively elaborated on the necessity to include diverse experts, and concluded that negotiating parties should think 'outside the box' and reach for experts far outside diplomatic circles.[38] For Borrie and Thornton, this does not only mean the inclusion of international organisations, civil societies or victim associations, but also the establishment of diverse negotiation formats

38 Borrie and Thornton, op. cit.

outside conventional diplomatic procedures.[39] The findings of the present chapter imply that both negotiations and the implementation of disarmament regimes may require diversity. In the case of the BTWC and the prohibition of offensive uses of biological agents, such diversity could be provided by the deepened exchange between the life-science and disarmament communities. Indeed, classrooms where graduate students, fireguards, military intelligence and disarmament experts meet seem to be unusual. Yet whether implemented metaphorically or in reality, such classroom settings might contribute to the necessary diversity and formulation of new long-term approaches that help to create the missing link.

39 Ibid. pp. 71–5.

Chapter 4: Israel

DAVID FRIEDMAN

This chapter examines recent activities in Israel to promote awareness and action in relation to biosecurity. 'Biosecurity' here refers to the sum total of measures aimed at preventing deliberate attempts to obtain dangerous biological agents or technologies and information that will grant the capability to make biological weapons. In other words, all the steps that must be taken to deny access by unauthorised actors to dangerous biological agents, information and technology that can be used to manufacture bioweapons.

Israel is an important country for examination in this regard. As the 2006 Lemon-Relman Committee Report by the US National Research Council[1] noted, almost 60 per cent of Israeli-authored scientific publications are in the life sciences, including medicine and the agricultural sciences. This report also indicated that the impact of citation of scientific research to the gross national product (GNP) scores highly in Israel compared to life-science industries in some 30 other global competitive countries. This means that scientific research is a key feature of Israel's GNP and 60 per cent of that research is related to the life sciences.

Much of this chapter focuses on the activities associated with the recently formed Steering Committee on Biotechnology Research in an Age of Terrorism (COBRAT). However, prior to that it is important to detail the context of attention to biological weapons in Israel along with its research system.

Background: Combating the Threat from Biological Weapons

To combat the bioterror threat effectively, a multi-system strategy is essential. Such a comprehensive plan must address prevention, defence, and consequence management. The objective of prevention is to stop or limit hostile forces from obtaining, developing, producing or using biological weapons. To prevent

1 Institute of Medicine and National Research Council 2006, *Globalization, biosecurity, and the future of the life sciences*, Washington, DC: National Academies Press.

states with developed scientific and technological infrastructures from attaining and manufacturing bioweapons if they choose to is almost impossible, although sometimes they can be deterred from using them. On the other hand, although it is difficult, it should be possible to prevent terrorist organisations from acquiring bioweapons, especially the more sophisticated, advanced and dangerous ones. However, this would require global cooperation, something not forthcoming when the terrorist organisation has a national sponsor or purveyor.

Traditionally, the majority of resources have been invested in defence, a strategy composed of protection, detection and early warning. When the main threats were from hostile states, this was justified. In order to design, develop and acquire an effective defence system, it was necessary to have accurate intelligence data concerning the enemy's plans, which was possible, albeit difficult. In contrast, it is almost impossible to predict the exact scenario of a bioterror attack. Therefore, defence systems may not give an optimal response when a strike occurs.

The aim of consequence management is to treat and save the lives of mass casualties. The basic building blocks of this goal are mainly medical measures, decontamination procedures, quarantine and evacuation. The source of an attack (terrorist or hostile state) is irrelevant. The only significant parameters are the number of casualties and the nature of the disease. Therefore, a country that is well prepared for a state-based biological threat will also be prepared for a bioterror attack. Moreover, since there is a great similarity between a bioattack and a natural epidemic, the most cost-effective approach is a 'dual-use' medical system, where the national medical setup is prepared for both cases.

Israel has had 50 years of experience in fighting conventional terrorism of various kinds. For most of that time, it also has been living under the shadow of a very real chemical and biothreat from many of its neighbours (for example, Syria, Iraq and Iran).[2] Over the years, Israel has developed very good defence and public-health (consequence-management) systems. It should be emphasised that the chemical and bioweapon threats are not only military ones; they are also a concrete threat to the Israeli civilian population.

When the biothreat re-emerged as an issue at the end of the 1990s, Israel recruited all its know-how and resources to modify its existing defence systems to include the new scenario. More recently, Israel has also begun to increase its emphasis on, and activity in, prevention.

---

2 Tucker, J. B. 2006, *War of nerves: chemical warfare from World War I to Al-Qaeda*, New York: Pantheon Books; Ali, J. 2001, 'Chemical weapons and the Iran-Iraq war: a case study in non-compliance', *Nonproliferation Review, CNS*, vol. 8(1).

Recent years have seen a revolution in the biological sciences. New molecular biological approaches and related technologies promise great benefits, but could also create more virulent micro-organisms that are resistant to antibiotics and vaccines, or that have other characteristics of effective biological weapons.[3] Initially, such new technologies may only be available to a select few, but the rapid dissemination of information through modern communications makes it possible for hostile forces to access them. Such forces can use them to develop and produce sophisticated, dangerous biological weaponry that would be very difficult to counter.[4] Thus, it is imperative to prevent organisms, knowledge and materials relevant to the production of bioweapons from reaching hostile hands.

Israel conducts world-class biomedical research. This is done in a number of sectors — at universities, research institutes, hospitals and government laboratories. A 2003 Israel National Security Council (INSC) survey performed by the Center for Technological Analysis and Forecasting (ICTAF, Tel Aviv University) identified close to 500 Israeli focal points of biological research, development, and manufacture of potential relevance to biological weapons. The analysis includes academic and non-academic research institutions, government organisations (for example, the Ministries of Health, Agriculture, and Science), and industry.

Work on micro-organisms, often virulent ones, takes place in about 50–100 laboratories. Most use advanced biological methods and technologies, and possess considerable manufacturing expertise and knowledge that is potentially relevant to developing bioweapons.

Israel's Biomedical Research and Development System

Organisationally and functionally, the system is extremely decentralised, with no single national authority having comprehensive responsibility for these laboratories and focal points. Instead, accountability is divided between a number of government ministries, authorities and academic institutions. No sole centralised authority deals formally with professional issues relevant to the

3 Chyba, C. F. and Greminger, A. L. 2004, 'Biotechnology and bioterrorism: an unprecedented world', *Survival*, vol. 46(2), pp. 143–62.

4 The proliferation of chemical and biological weapons, materials and technologies to state and sub-state actors 2001, testimony by Jonathon B. Tucker, before the Senate Subcommittee on International Security, Proliferation and Federal Service, available: http://cns.miis.edu/research/cbw/cbwol.html; Pate, J. and Ackerman, G. 2001, 'Assessing the threat of mass-casualty bioterrorism', Nuclear Threat Initiative, available: www.NTI.org/e_research/e3_1b.html.

proper performance of this system, and neither does any national organisation possess complete information about the system's scientific personnel, their research interests and their laboratory's research.

Ministry of Industry, Trade and Labour (MITL)

De jure, the primary legislated responsibility for worker and workplace safety, and hence laboratory biosafety, rests with the MITL. *De facto*, oversight and supervision of Israel's biomedical laboratories is considerably more complicated. The MITL tends not to focus its attention, expertise and inspections in the science sector. In contrast, the Ministry of Health (MOH) has major, expanding interest, expertise and, increasingly, activity in the field, which falls under the rubric of its general mandate to promote national health (see later in chapter).

The MITL's Laboratory Accreditation Authority (LAA) was established by law to accredit and inspect laboratories and ensure their compliance with international quality and safety standards. Observance of each measure is judged separately; there is no evaluation of the laboratory as a whole. Israeli law requires LAA accreditation only in specific sectors; for example, cement standards. In all non-specified areas, it is voluntary. This can lead to some unevenness. For example, the MOH's water and food laboratories must be certified, but the same ministry's medical laboratories are exempt. A few Israeli medical laboratories do seek voluntary endorsement for commercial reasons; but the lack of a comprehensive accreditation requirement for all biomedical laboratories prevents their effective central regulation.

Ministry of Health (MOH)

The MOH's responsibilities vary for Israeli biomedical laboratories in different sectors. Most conduct research and undertake routine diagnoses, are situated in hospitals, and many work with virulent bacteria or viruses. Laboratories in state-owned hospitals are under full MOH supervision. Other hospitals and laboratories belong to one of Israel's Kupot Holim (private health plans [HMOs]). These are not under MOH supervision, direct or indirect. The country's medical schools enjoy absolute independence and are not supervised by the MOH; rather, each medical school/university has its own safety committee.

The MOH's own Department for Laboratories, part of the Public Health Service, is directly and fully responsible for the operation of the ministry's six internal public-health laboratories. It also provides varying amounts of administrative oversight for hospitals, public- and private-sector medical laboratories and LAA-accredited environmental-health laboratories.

The MOH must approve medical laboratories in hospitals and HMOs and their professional staff. However, the ministry's Department of Laboratories does not possess information about, much less oversee, the research actually carried out in such environments. A dwindling number of private medical laboratories (only 13 are still operating today) are, in principle, supervised by the MOH.

Recently, the MOH has begun to expand its oversight of medical laboratories in hospitals and health-service organisations, including the tracking of biological agents, the registration of workers, and, for the last few years, a regime of regular inspections. The Ministry's six public-health laboratories follow orderly safety procedures, including registration and documentation. However, the MITL biosafety regulations provide oversight and supervision of all other medical-laboratory work. The MITL regulations assign broader responsibility for these issues to the laboratory director, who must also appoint a safety supervisor.

Other (often industrial) private laboratories are not classified as 'medical laboratories', but as 'biological laboratories', although they do work with dangerous biological agents. The MOH does not oversee these laboratories in any way. Such laboratories just need a MITL business licence and are subject only to the usual MITL biological-safety oversight. This potentially serious problem should be tackled within some appropriate framework.

Institutions of Higher Education

The lion's share of Israeli life-sciences and medical research and development is conducted at the country's universities and academic-research institutions: the Hebrew University, Tel Aviv University, Ben-Gurion University, Bar-Ilan University, the University of Haifa, the Technion, and the Weizmann Institute of Science.

Israel's universities are not formally subordinate to any government or public body, although they retain strong links to the Council for Higher Education and its Planning and Budget Committee that divides the government's total budget for higher education among them. All Israeli institutions of higher education share a similar organisational structure — a president, who usually appoints a vice-president for research and development, heads each.

Individual university scientists usually enjoy considerable scientific freedom with no institutional reporting, oversight or supervision. Their work is only reviewed once every few years in the framework of institutional promotion committees. Only a few special activities are regulated by national or organisational procedures. For example, an Animal Experimentation Law establishes standards for the use of research animals; and Helsinki Committees oversee experiments on humans. Work with dangerous biological agents and poisons are regulated under Israel's extensive biosafety legal infrastructure, and academic establishments have appropriate procedures and organisations to ensure compliance (dangerous biological agents, defined as micro-organisms and toxins that cause disease in humans, are listed in the law).

The awareness of biosafety and its legal requirements is increasing. Since international research funding bodies (for example, the US National Institutes of Health and the US military) are demanding more effective biosafety supervision and oversight in the foreign laboratories they support, Israel's academic biosafety procedures are continuously improving. All educational research institutions have safety units, a full-time safety director, and safety committees. Each safety system complies with the relevant laws and directives of the MITL Workplace Inspection Division. Appropriate laws include the Workplace Safety Order (1970), the Workplace Inspection Organisation Law (1945), and the Safety Oversight Order for Medical, Biological and Chemical Laboratories (2001).

An institution's safety officials oversee work with dangerous biological agents as listed in the law, with human blood and tissue samples, DNA manipulation, toxic materials, and pathogenic organisms. Workplace regulations and guidelines are constantly updated, and laboratories are inspected regularly to ensure compliance. Record-keeping and periodic reporting regarding high-risk materials are required, and automated systems are being created to track the purchase of dangerous strains and special biological materials.

Biosafety oversight in academia takes place at two loci: first, when research proposals are submitted, and second, during its progress. In addition, safety authorities conduct instructional workshops for scientists, laboratory workers and students in safety procedures. In some institutions, when a research project requires safety certification, it is given only after the safety division has confirmed that the laboratory's work conditions meet legal requirements.

Biosecurity in Israel

Israel, the US and Western Europe share common views concerning the threat of bioweapons, bioterror and the creation of weapons of mass destruction (WMD) among rogue states and terror organisations.[5] Israel has repeatedly stated that its national policy is to prevent such proliferation, and has taken concrete steps in this direction, some in the framework of internal legislation and some as part of international initiatives, including those of the UN.

---

5 Danzig, R. 2003, *Catastrophic bioterrorism: what is to be done*? Center for Technology and National Security Policy, National Defense University, Washington, DC.

Although Israel has not formally joined the Biological and Toxin Weapons Convention (BTWC), regarding it as an inseparable part of a general and comprehensive regional political arrangement, it wholeheartedly adheres to the US, EU and UN initiatives combating bioterror and WMD proliferation. It adheres to and coordinates its activities with the AG regime and fully supports UN Resolution 1540. Israel also has consistently supported the policy of the US in its war against international terror of all kinds.

However, unlike the US, Western Europe and other countries, Israel has yet to adopt legislation directly aimed at preventing or minimising the spread from its own laboratories, of non-conventional weaponry and its components, including dangerous biological agents.

While, as outlined in the previous section, Israel has a well-developed system of civilian biosafety (as distinct from biosecurity) laws and regulations, these can make only a limited and indirect contribution to oversight and inspection aimed at preventing the seepage of dangerous agents or information into hostile hands. There is also an executive order issued by the MITL in 2004 which mandates the oversight of chemical, biological and nuclear exports 'to help prevent the spread of non-conventional weaponry... [by] forbidding the export from Israel of products, technologies and services that can be used to develop and manufacture chemical, biological or nuclear weapons'. It is important to note that to minimise any harm to basic and clinical biomedical research this MITL order specifically exempts the export of chemical and biological agents used for medical and veterinary diagnosis, treatment or research, and information related to such agents.

The prevention of biological terror remains of supreme importance at national level. A preliminary study at the INSC in 2003 produced the following findings:

- There is virtually no awareness of the need for biosecurity within Israel's civilian life-sciences research community.
- Israel has no legal and/or regulatory infrastructure directed specifically towards biosecurity. Existing biosafety laws and regulations provide only indirect and partial means for dealing with biosecurity.
- Institutions where biomedical research and development and other work (diagnosis, production, and so on) is performed are not subject to inspection or supervision by any single Israeli authority or ministry. Instead, this responsibility is shared between a number of ministries, where division of responsibility is often not clear.
- As a result, neither at national or ministerial level is there a system of control or supervision of biomedical research laboratories, nor is there sufficient

information about any dangerous biological agents used, the types of research performed, or the technologies employed.

The COBRAT Report and its Recommendations

The big challenge now is to incorporate biosecurity concerns into the system, in particular, to upgrade measures to prevent the leakage of dangerous organisms, information and technologies to terror organisations. To this end, the INSC and the Israel Academy of Sciences and Humanities (IASH) initiated a national project called 'Biotechnology Research in an Age of Terrorism', and formed a special Steering Committee on Biotechnology Research in an Age of Terrorism (COBRAT) to analyse and report on the current situation and recommend future action.[6] The committee was composed of well-known scientists and biologists from Israeli academia and industry and experts in regulatory and legislative law, similar to the Fink Committee in the US.[7]

COBRAT took the situation in which authorities and the scientific community are oblivious to biosecurity issues as its starting point in seeking more effective and systematic ways to meet biosecurity concerns without compromising academic freedom and creativity. In its final report the Committee formulated specific recommendations to address:

- the changes required in Israel's existing legislative infrastructure
- the compilation of an updatable list of biological agents and research topics requiring inspection and supervision
- the establishment of a regime for tracking, supervising and enforcing all areas of biosecurity
- the need for a national inter-ministerial body or professional committee to guide, monitor and maintain biosecurity.

In pursuing these goals, COBRAT was confronted by several daunting but not atypical facts: (1) no biosecurity legislation exists in Israel; (2) the legislative process, as practiced by the Israeli parliament (Knesset), is long, complicated and uncertain; (3) a response to the bioterror threat cannot wait for long-term solutions. COBRAT's innovative yet practical interim solution to these problems may also serve as a useful model for others. COBRAT recommended modifying Israel's biosafety committees and empowering them, by executive order, to undertake responsibility for biosecurity concerns as well. In addition to reducing duplication, disruption and delay, this scheme avoids many of the

6 Steering Committee on Issues in Biotechnological Research in an Age of Terrorism 2008, Report by the Israeli Academy of Sciences and Humanities and the Israeli National Security Council.

7 National Research Council 2004, *Biotechnology research in an age of terrorism*, Washington, DC: National Academies Press.

sensitivities, suspicions and conflicts inherent in the regulation of dual-use research. The existing biosafety committees are of long standing, sensitive to scientific concerns (and those of the individual scientist), well-tolerated by the scientific and academic communities, and unlikely to trigger the hostility and 'graft rejection' typical of introducing a 'foreign body' into academia. Trust and comfort are indefinable, but their effects are all too real.

With this introduction let us proceed to the Committee's (edited) recommendations given in Table 1.[8]

Table 1: COBRAT's Recommendations

Recommendation 1: awareness, consciousness and education

An ongoing effort should be carried out to raise awareness and understanding of the risks associated with the biological threat in general, and with dual-use biological research in particular, within Israel's life and medical research and development community.

Recommendation 2: existing and new legislation

Legislative solutions must be addressed on two levels:

Since the creation of totally new legislation, under Israeli conditions, can be a long, slow and uncertain process, the Committee recommends that existing Israeli secondary legislation on biosafety should immediately be used as a model for ministerial executive orders and institutional (for example, university) procedures designed to prevent the potential seepage of organisms, materials and information to hostile elements.

In parallel, specific longer-term legislation should be formulated. This legislation must be comprehensive and cover all aspects of biosecurity.

Recommendation 3: oversight and supervision mechanisms

The fastest, most efficient and least disruptive way to enforce a regime ensuring biosecurity is to upgrade and adapt existing institutional biosafety oversight procedures to also assure biosecurity.

8 Friedman, D., Rager-Zisman, B., Bibi, E. and Keinan, A. 2008, 'The bioterrorism threat and dual-use biotechnological research: an Israeli perspective', *Science and Engineering Ethics* vol. 16(1), pp. 85–97.

Local responsibility for the enforcement should be delegated to existing institutional biosafety committees (renamed "biosafety and biosecurity committees") for the academic sector and special Central Biosafety and Biosecurity Committees for biomedical laboratories affiliated with government ministries. National biosecurity policy, procedures and enforcement should be overseen by a National Biosecurity Council (NBC) to be appointed by the Ministry of Health (MOH).

Recommendation 4: list of dangerous agents

There should be an itemized core list of dangerous agents. Not all biological agents should be placed in this category. The list of agents issued by the US Department of Health and Human Services was adopted as the initial core list. The list should be reviewed and updated annually, as required, by the NBC. The Committee emphasises, however, that sensitive dual-use data and information are not limited to research connected with these agents, but also can stem from work with other, in themselves harmless, strains.

Recommendation 5: publication of information generated by dual-use research

This sensitive subject must be an essential part of Israel's biosecurity policy. Given the risks involved, it is recommended to establish a system to oversee and approve the publication of the results of dual-use research projects. This should be undertaken by an internal mechanism based on the judgment of the academic community itself. Professionalism, balance and lack of undue delay will be essential to ensuring acceptance.

Recommendation 6: consideration of biosecurity issues by funding agencies

It is recommended that the Israel Science Foundation (ISF) and government research foundations require, as part of their approval process, biosecurity approval from the applicant's institution. This would ensure that these issues are considered by applicant institutions and that proper safety and security measures are enforced. In the case of non-academic laboratory research, similar certification should come from the chairman of the Central Safety and Security Committee in the relevant ministry.

Recommendation 7: supervision of importation and sale of dual-use biological equipment and agents

In addition to existing export regulations, the Committee believes that it is necessary to establish a system to oversee the Israeli import of dual-use biological laboratory equipment and biological agents, as defined by the (export) risk list maintained by the MITL Export Authority, as well as the sale of these items in the local market (in particular, the sale of used equipment).

Recommendation 8: national responsibility for biosecurity

The establishment of a biosecurity regime and its enforcement should be assigned to the Ministry of Health (MOH), which has both primary responsibility for public health and the requisite scientific knowledge and professional experience. MOH should establish a National Biosecurity Council (NBC). The Chairman and members of the Council should be appointed by the Minister of Health in consultation with the head of the National Security Council and the president of the Israel Academy of Sciences and Humanities.

The New Legislation Process

In its work the Committee has sought to clarify the extent to which Israeli law contains normative instructions to deal with bioterror threats that could result from scientific research conducted in Israel's biological and medical laboratories. The Committee found that Israel lacks legislation specifically addressing this goal, although there are many relevant existing statutes. In particular, there is a clear link between the need to protect the safety and health of laboratory workers handling dangerous biological agents and the public at large. Thus the Committee carefully examined existing biosafety laws that address inspection, work safety, hygiene and public health as they relate to biological laboratories.

The Committee has concluded that, although Israel has an effective legal framework for biosafety, it urgently needs a similar normative structure for biosecurity. A statutory list of dangerous biological agents and their forbidden uses must be drawn up and updated frequently. Relevant laboratories must be identified and certification procedures for using dangerous organisms legislated. Legislation must also provide for the adequate supervision of anti-theft, transfer and storage procedures. Clearly, existing biosafety provisions intended to protect people working with dangerous biological agents from laboratory accidents are also relevant for biosecurity.

Other efforts must include raising the awareness of laboratory directors, scientists and students regarding existing legal requirements, the current bioterror threat, and the vital need for biosecurity and biosafety procedures. An active concern for biosecurity plays an important role in establishing standards for working with dangerous biological agents. International initiatives followed by national legislation in many states focus on laboratories holding stores of dangerous biological agents, because these are prime targets for hostile forces.

'Preventative caution' requires rules that specify how to prevent hostile forces from acquiring bioweapons. The Committee believes that any framework must provide for the continued performance, publication and implementation of scientific research, as well as the defence, oversight and inspection mechanisms needed to prevent or minimise hostile use of ostensibly positive research results.

The committee assumed that introducing a new law would be a lengthy process and therefore recommended an interim step be taken. This step was to integrate biosecurity laws into the existing biosafety laws and regulations. Fortunately, and contrary to the committee's expectations, a separate biosecurity law was put on fast track, thanks to the combined efforts of several members of parliament.

The Regulation of Research into Biological Disease Agents Act, 2008

In November 2008, the Israeli parliament passed legislation on a set of laws that cover biosecurity issues. Moreover, the main recommendations of the committee were made law thanks to their cooperation with the different government departments, mainly the MOH and Ministry of Justice.

The main points of the law are as follows:

- The law applies to all institutions and laboratories (universities, research labs, industry and hospitals), in all sectors, that have in their possession disease-causing biological agents as listed in the law or conduct research or diagnostics in said agents.
- The Minister of Health will be in charge of enforcing this law in all institutions.
- Possessing, conducting research or working with these biological agents requires an authorisation from the Ministry of Health.
- Possessing, conducting research or working with these biological agents must be performed so as not to impinge upon safety or security concerns.
- No one shall conduct research whose sole purpose is to cause or exacerbate a disease or illness or to impair the ability to prevent or treat it.
- A person or institution that has conducted a research study for which permission did not have to be obtained under the Act, but which has made

findings of a nature to increase the virulence or the contagiousness of disease agents not included in the list, or findings of a nature to alter the host range of the said disease agents, so that the disease can pass to humans, shall halt the research and submit a request to the 'institutional committee'.

- All institutions that possess disease-causing agents will establish an 'institutional committee' whose purpose is to supervise the research conducted in that institution. The committees will comprise scientists as well as security and safety personnel from the institution.
- A council for biological disease-agent research will be created to advise the Minister of Health and will comprise professionals and members of relevant government ministries. The council's responsibilities will be to advise the Minister of Health regarding research authorisation, to supervise the various institutions, and to promote training workshops and courses in institutions that work with biological agents.

Since the Act itself does not define what 'public information campaigns' and 'in-service training courses' are, it is understood that the Council has a duty to oversee those through to their implementation. Based on the Act, the Council can approve operating rules that are implemented by institutional committees (of a corporation or company conducting research, whether scientific, medical, industrial-commercial or educational, including hospitals and government organisations) to approve scientific research in Israel. Therefore, the Council can guide research establishments to adopt such campaigns or training courses as a part of their operating rules, notably so for educational institutions. Moreover, the Council also has the right to oversee the institutional committees' compliance with their operating rules based on the Act. Once certain educational programmes are set out, research establishments need to be compliant with them. Education of life scientists about dual use in Israel will be an important case where specific national legislation to deal with biosecurity is achieved alongside the establishment of specifically dedicated committees to address biosecurity issues.

The authorisation process allowing institutions to posses and/or conduct research with biological-disease agents was launched in Israel in 2009 and organisations have begun establishing their own internal committees. Towards this end, a nationwide workshop is planned to take place at the end of February 2010. Participants will include members from the Council and institutional committees, plus other representatives from various establishments. The current legislation and regulations will be discussed and clarified and the participants will also hear lectures on biological threats, dual-use purpose research and more.

Awareness-raising and Education

The recent Regulation of Research into Biological Disease Agents Act is without a doubt a giant step forward in Israel's awareness and attitude towards biosecurity. Nevertheless, we should anticipate a long and gradual process that will require a great deal of effort and patience. The success of this legislation is largely dependent on researchers and their cooperation is crucial. For biosecurity regulation to succeed, researchers must first be well-informed about the topic. It is important that they recognise and understand the potential harm that can be caused by the technologies they are developing and the research they are conducting.

In this regard, it is important to note that raising awareness about biosecurity has already received some attention in Israel in recent years. Numerous programmes have been launched to assess the level of knowledge among different research communities, as well as offer ways of increasing awareness. Two major figures in this move are Malcolm Dando and Brian Rappert, [9]who have been very active in launching programmes and publishing a large number of papers and books on the topic. Such a programme is currently running in Israel, with the support of the Sloan Foundation. The initial stage of this programme — which included a survey conducted by the author — investigated the relevance of courses in bioethics, biosecurity and biosafety within Israel's research universities. [10]

The survey examined the syllabi of 35 courses offered at the Faculty of Life Sciences of six research universities in Israel. Courses were sampled by focusing on those that provide specific educational modules on biosafety, biosecurity, and bioethics. The rationale for the survey was that we aimed to investigate the current state of awareness regarding these topics within the research communities of life sciences, as it manifests itself in the curricula.

In general, we found that very little biosecurity education is offered to researchers in the life sciences. Moreover, the results indicate there is currently no academic course at Israeli universities that is specifically designed to educate life scientists on the issue of biosecurity. Interestingly, comparable surveys conducted in different parts of the world have rendered very similar results (see the chapters in this volume by Minehata and Shinomiya, and Mancini and Revill).

---

9 Rappert, B. 2007, 'Education for the life sciences', in Rappert, B. and Mcleish, C. (eds), *A Web of prevention: Biological weapons, life sciences and the future governance of research*, London: Earthscan; Rappert, B., Chavrier, M. I. and Dando M. R. 2006, *In-depth Implementation of the BTWC: Education and Outreach*, Bradford Review Conference Papers, no. 18.

10 Minehata, M. and Friedman, D. 2010, *Biosecurity education in Israeli research universities: Survey report*, Bradford Disarmament Research Centre (BDRC) and Institute for National Security Studies (INSS).

Given these results, it would be reasonable to assume that insufficient education is a major contributing factor to the lack of awareness of biosecurity issues amongst life scientists. In a similar vein, the COBRAT Report also noted the lack of legal infrastructure for biosecurity and the fact that there is virtually no knowledge of the need for biosecurity amongst Israel's scientific-research community as well as within Israeli civilian life. What is more, this lack of awareness was concluded to be the likely reason for biosecurity education also being essentially non-existent in Europe and Japan.

Nonetheless, it is important to note that even universities that were informed about biosecurity issues, in Israel as well as Europe and Japan, still encounter difficulties when attempting to include biosecurity education in their curricula. These difficulties may include:

- insufficient time available in the existing curricula
- time constraints and insufficient resources required for the development of new curricula
- lack of expertise and available literature on biosecurity education
- lack of interest in biosecurity education.

With the results of the survey in mind, the second part of the Sloan Foundation-supported programme aims to raise awareness. In this part of the programme, 10 life-science faculties and/or departments at research universities in Israel (Tel Aviv University, Ben-Gurion University, the Technion, Bar-Ilan University and the Hebrew University) were targeted and an hour-long seminar was given in each on the subject of dual-use research and biosecurity. The audience comprised faculty members as well as graduate students. The seminar discussed the threat of bioterrorism, the potential dangers posed by advanced biotechnological research and the possible systems that can be implemented to stop or greatly hinder the transfer of biohazardous material and sensitive information into the hands of terrorists. In addition, the new Israeli legislation was presented and discussed. Following the lectures, a questionnaire was sent to all who had attended. These questionnaires will be analysed and used to determine the programme's next steps.

This series of seminars, limited as it may be, is nevertheless an important first step towards increasing awareness regarding biosecurity issues. Therefore, one of the main goals of the programme, which will be based on the analysis of the questionnaires, is to build a lesson plan or course syllabus on the subject. With the help of the Council for Biological Disease Agent Research and the MOH, we will encourage research institutions that deal with biological agents to incorporate such courses in their curriculum. With this, we hope to contribute considerably to biosecurity education and expect to see a significant rise in knowledge of biosecurity issues in Israel.

In the context of raising awareness, it is important to discuss the above-mentioned Regulation of Research into Biological Disease Agents Act, which was recently passed in Israel, as well as the Council for Biological Disease Agent Research, which was established under this Act. Although Israel is not a State Party to the BTWC, this legislation certainly conforms to the spirit of the treaty and puts Israel at the forefront of confronting the issues of biosecurity. In this sense, the education of life scientists about dual use in Israel will be an important test for examining the effects of national legislation on biosecurity and national committees specifically targeting biosecurity issues.

To sum up, based on the data presented above, this investigation indicates that there is a lack of biosecurity education and educational content on dual-use issues in Israel at the time of inquiry. However, this certainly does not mean that promoting biosecurity education in Israel cannot be done. In fact, we believe precisely the opposite is the case, perhaps most importantly as a result of the Regulation of Research into Biological Disease Agents Act and the establishment of the Council for Biological Disease Agent Research. The Council is responsible for outlining, recommending and overseeing the implementation of regulations enhancing biosecurity at research institutions in Israel alongside raising awareness of biosecurity issues amongst life scientists. Hence, the Israeli government's initiative to develop infrastructure for biosecurity policies is evident in this Act and, in this sense, is an example of a top-down approach to the promotion of biosecurity education whereby raising awareness begins with legislation, then trickles down to the level of educational institutions and, finally, reaches the public sphere.

Summary and Conclusion

Over the past 40 years, the state of Israel has been facing chemical and biological threats, not only to its military but also mainly to its civilian population. Until the late 1990s, the threat emanated primarily from hostile states that developed and stockpiled bioweapons and chemical weapons. However, from that time, and especially after the 11 September attacks and subsequent distribution of anthrax envelopes around the US, bioterror has become a global threat. Israel joined the international effort spearheaded by the US to curb this threat, investing much of its resources on building an effective biodefence system, as well as joining efforts to prevent bioweapons and their components from reaching hostile hands, and mostly stopping the leakage of dangerous biological agents and dual-use technologies and information to terrorists.

In order to assess and investigate the issues in Israel, the COBRAT was assembled. The committee's main recommendations were: 1) to initiate and enhance education and awareness in the life-science community; 2) to promote legislation; and 3) to establish a regulation system regarding research with dangerous biological agents and dual-use technologies. These proposals were quickly implemented, as evidenced by the establishment of the Council for Dangerous Biological Agents as well as novel legislation, placing Israel at the forefront of countries confronting these issues successfully.

The following list summarises the most important and effective properties of the Israeli approach to the fight against bioterror:

- a top-down approach, whereby official agencies initiate assessment and research of the issues, leading to legislation, which is subsequently followed by structured education at university level, and finally, the launching of public campaigns
- the assessment and research is independent and conducted by senior scientists from the academic life-science community, rather than government officials
- cooperation between the INSC, which represents the interests of national security, and the Israeli Academy of Science, which stands for pure academic research
- the support of public officials, such as members of Parliament
- the establishment of a professional advisory council that is responsible for implementation and supervision of the law
- the legislation regulates not only research with specifically listed biological-disease agents but also dual-use research
- although research institutes are regulated by the Council for Dangerous Biological Agents they each have the mandate to work as an independent entity; a configuration which significantly reduces bureaucracy.

In conclusion, we strongly believe that the Israeli approach for addressing and confronting current biosecurity issues and, in particular, the top-down approach, is the optimal model and should be adopted by governments across the world.

Chapter 5: Japan: Obstacles, Lessons and Future

MASAMICHI MINEHATA AND NARIYOSHI SHINOMIYA

Japan has a clear rationale to discuss the introduction of ethical education for life scientists regarding its dual-use dimensions.[1] This partly derives from the size of its life-science industry and from the actual threats posed by the misuse of science. Japan has been one of the leading global marketplaces of the life-science industry.[2] This indicates that a large number of life scientists are practising cutting-edge research in Japan. Importantly, some of them have misused their knowledge in the form of biocrimes and bioterrorism. One of the most prominent cases of such misuse was that of the religious group Aum Shinrikyo. By recruiting scientists from top academic institutions, the group was able to conduct sarin attacks on the Tokyo subway in 1995. The group also attempted several biological attacks using botulinum toxin and anthrax from 1990 to 1995.[3] Therefore, enhancing ethical awareness among scientists is of critical importance in extending their moral responsibility to do no harm and minimise any potential damage to humans, animals and plants.

Although attempts to define 'biosecurity' are not straightforward,[4] in this chapter it is conceptualised as taking both 'preventative and responsive measures, in a multifaceted manner, to mitigate the multidimensional threat

1 In this chapter, dual-use refers to the possibility whereby peacefully developed scientific research can be applied for malign purposes, such as biowarfare and bioterrorism.

2 See Chapter 3 of National Research Council 2006, *Globalization, biosecurity and the future of the life sciences*, Washington, DC: National Academies Press.

3 Sugishima, M. 2003, 'Biocrimes in Japan', in Sugishima, M. (ed.), *A comprehensive study on bioterrorism*, Legal Research Institute Monograph, Japan: Asahi University; Wheelis, M. and Sugishima, M. 2006, 'Terrorist use of biological weapons', in Wheelis. M., Rozsa, L. and Dando, M. R. (eds), *Deadly cultures: Biological weapons since1945*, MA: Harvard University Press; Takahashi, H., Keim, P., Kaufmann, A. F., Keys, C., Smith, K. L., Taniguchi, K., Inouye, S. and Kurata, T. 2004, 'Historical review: Bacillus anthracis incident, Kameido, Tokyo, 1993', *Emerging Infectious Diseases*, vol. 1(1), pp. 117–20.

4 The term 'biosecurity' has been defined in different concepts in different social and linguistic backgrounds in different countries. See Sunshine Project 2003, 'Biosafety, biosecurity, and biological weapons', background paper on three agreements on biotechnology, health, and the environment, and their potential contribution to biological weapons control, October 2003, available: http://www.natwiss.de/publikationen/Biosafety_and_Biosecurity.pdf [viewed 17 January 2010]; Fidler, D. and Gostin, L.L. 2007, *Biosecurity in the global age: Biological weapons, public health, and the rule of law*, CA: Stanford University Press.

posed by bioterrorism, biowarfare, and the potential advertent or inadvertent misuse of the life sciences'.[5] This is a broader concept than that often given to 'laboratory biosecurity',[6] although in Japan at this time the latter is generally understood as 'biosecurity'.[7] Therefore, in this chapter biosecurity education is also widely envisaged as a process to better inform understanding of how the possible misuse of the life sciences can be prevented. It includes themes such as, *inter alia*, the history of state-level offensive biological-warfare programmes and biological terrorism; the history and evolution of the international prohibition regimes and their national implementation;[8] dual-use risks and ethical responsibilities of life scientists; and building an effective set of preventative policies to ensure benign development of the life sciences.

In order to improve biosecurity education in Japan, it is necessary to understand the existing provisions regarding dual-use issues and learn lessons from the implementation of such teaching. This is useful for accumulating knowledge in biosecurity education to share with other countries. To achieve this purpose, this chapter will give a brief overview of Japan's stance towards international efforts to mitigate the threat posed by misuse of the life sciences. Secondly, the chapter shifts its focus to a domestic context by providing the survey results on biosecurity educational provisions in 197 university-level life-science degree courses. Thirdly, biosecurity education at Japan's National Defense Medical College (NDMC) will be used as an example of the introduction of such education. Finally, the way forward for the education of life scientists in Japan will be envisaged.

---

5 Minehata, M. and Shinomiya, N. 2009, *Biosecurity education: enhancing ethics, securing life and promoting science: dual use education in life-science degree courses at universities in Japan*, Saitama and Bradford: National Defense Medical College and the University of Bradford, available: http://www.dual-usebioethics.net/ [viewed 17 January 2010]. There have been efforts to conceptualise a multifaceted approach comprising several practical measures through what is termed the Web of Prevention (WoP). For the conceptual evolution of the WoP in literature, see Feaks, D., Rappert, B. and McLeish, C. 2007, 'Introduction: A web of prevention', in Rappert, B. and McLeish, C. (eds), *A web of prevention: Biological weapons, life science and the governance of research*, London: Earthscan.

6 The World Health Organisation definition of laboratory biosecurity refers to 'institutional and personal security measures designed to prevent the loss, theft, misuse, diversion or intentional release of pathogens and toxins'. See World Health Organisation 2004, *Laboratory Biosafety Manual*, Geneva: WHO, p. 47.

7 Furukawa, K. 2009, 'Dealing with the dual-use aspects of life science activities in Japan', in Rappert, B. and Gould, C. (eds), *Biosecurity: Origins, transformations and practices*, Hampshire: Palgrave Macmillan.

8 Such as BTWC of 1972, Chemical Weapons Convention of 1993 or Geneva Protocol of 1925.

Japan's Stance on the Dual-Use Issue: The International Context

Amongst other calls to the international community,[9] States Parties of the Biological and Toxin Weapons Convention (BTWC) have in recent years conducted in-depth discussions on, for example, national measures to implement laboratory biosecurity (2003), the adoption of codes of conduct for scientists (2005) and the promotion of education on dual-use issues (2008).[10] As a result of these processes, Japan has been constantly considering its own views.

In the discussion on codes of conduct for life scientists, in particular, Japan addressed some key elements of awareness-raising among scientists. It explained that the lack of awareness among scientists was not to be taken as a sign of 'the immorality of scientists'; rather, 'the misconduct and failures of scientists are not caused by a lack of ethics but rather by ignorance'.[11] Therefore, the objective of the codes was the reduction of 'the risk of sciences causing negative effects on human beings and society through establishing specific rules that scientists should abide by'. Japan proposed to 'ensure scientists realize the risks of biological agents they handle, the possibility their research results may be abused and the effects of them actually being abused'. Furthermore, it was acknowledged that scientists themselves should be the 'core people' to formulate such codes, although involvement by other people concerned is also necessary.[12] At the same meeting the Japan Bioindustry Association (JBA) also illustrated its mandatory professional rules and guidelines, stating that such standards were important in ensuring both 'corporate compliance' and social responsibility of the industrial sector.[13]

9 The necessity of education wastouched upon in the *Statement on Biosecurity* by the InterAcademy Panel, which was endorsed by the national science academies of more than 60 states in 2005; see InterAcademy Panel 2005 *Statement on Biosecurity*, available: http://www.interacademies.net/?id=4909[viewed 17 December 2008]; World Health Organisation 2007, *Scientific working group on life science research and global health security: Report of the first meeting*, Geneva: WHO.

10 United Nations 2002, *Final Document of the Fifth Review Conference of the States Parties to the Convention on the Prohibition of the Development, Production and Stockpiling of Bacteriological (Biological) and Toxin Weapons and on Their Destruction*, BWC/CONF.V/17, Geneva: UN, available: http://www.opbw.org/ [viewed 17 January 2010]; United Nations (2006) *Final Document of the Sixth Review Conference of the States Parties to the Convention on the Prohibition of the Development, Production and Stockpiling of Bacteriological (Biological) and Toxin Weapons and on Their Destruction*, BWC/CONF.VI/6, Geneva: UN, available: http://www.opbw.org/ [viewed 17 January 2010].

11 Japan 2005a, *Codes of conduct for scientists: Discussions in Japan on the issue*, BWC/MSP2005/MX/WP.21, Geneva: UN, available: http://www.opbw.org/ [viewed 17 January 2010], p. 4.

12 .Ibid. p. 4.

13 Japan 2005b, *Codes of conduct for scientists: A view from analysis of the bioindustrial sectors in Japan*, BWC/MSP2005/MX/WP.22, Geneva: UN, available: http://www.opbw.org/ [viewed 17 January 2010], p. 3.

Lack of Awareness

Despite the amount of international attention given to awareness-raising among scientists, specific provisions for biosecurity education are not prevalent in many countries. This deficiency has been elaborated in the national papers of States Parties to the meetings of the BTWC, particularly by Australia,[14] the UK and the Netherlands.[15] Experts within the non-governmental community have also reached similar conclusions. For example, after conducting some 90 interactive seminars with more than 2500 life scientists in 13 different countries, Dando and Rappert concluded that there was a pervasive lack of awareness amongst individual scientists of the dual-use aspects of their research.[16] This was further supported by the survey on biosecurity education in European universities by Giulio Mancini and James Revill, demonstrating the deficiency of education in dual-use issues for life scientists.[17]

At a BTWC meeting in 2008, Japan acknowledged that 'the development of educational programmes at the government level has not seen great progress'.[18] One explanation for this is that, despite a growing attention to biosecurity issues and the development of some related regulations, human and financial resources to institutionalise and coordinate preventative measures to minimise dual-use risks in the life sciences are still limited amongst relevant ministries and scientific communities.[19] Thus it could be expected that biosecurity-education provisions have not been prevalent in higher-education institutions in Japan.

---

14 Australia 2005, *Raising awareness: Approaches and opportunities for outreach*, BWC/MSP/2005/MX/WP.29, Geneva: UN, available: http://www.opbw.org/ [viewed 17 January 2010].

15 UK and Netherlands 2005, *Oversight, education and awareness-raising: Report of a UK seminar*, BWC/MSP/2008/MX/WP.10, Geneva: UN, available: http://www.opbw.org/ [viewed 17 January 2010].

16 Dando, M. R. and Rappert, B. 2005, 'Codes of conduct for the life sciences: Some insights from UK academia', *Bradford Briefing Papers*, no. 16, available: http://www.brad.ac.uk/acad/sbtwc/briefing/BP_16_2ndseries.pdf [viewed 17 January 2010]; Rappert, B., Chevrier, M.I. and Dando, M.R. 2006, 'In-depth implementation of the BWC: Education and outreach', *Bradford Review Conference Papers*, no. 18, available: http://www.brad.ac.uk/acad/sbtwc/briefing/RCP_18.pdf [viewed 17 January 2010]; Rappert, B. (2009) Experimental secrets: International security, codes, and the future of research, New York: University Press of America.

17 The survey reported that only three out of 57 universities investigated in Europe offered specific biosecurity modules. See Mancini, G. and Revill, J. 2008, *Fostering the biosecurity norm: Biosecurity education for the next generation of life scientists*, Como and Bradford: Landau Network-Centro Volta and the University of Bradford, available: http://www.centrovolta.it/landau/2008/11/07/FosteringTheBiosecurityNormAnEducationalModuleForLifeSciencesStudents.aspx [viewed 17 January 2010]; Revill, J., Mancini, G., Minehata, M. and Shinomiya, N. 2009, 'Biosecurity education: Surveys from Europe and Japan', *Background paper for the international workshop on promoting education on dual-use issues in the life sciences*, 16-18 November 2009, Warsaw, Poland: Polish Academy of Sciences, available: http://dels.nas.edu/bls/warsaw/background.shtml [viewed 18 January 2010].

18 in consultation with Australia, Canada, Republic of Korea, Switzerland, Norway and New Zealand2008, *Oversight, education, awareness raising, and codes of conduct for preventing the misuse of bio-science and bio-technology*, BWC/MSP2008/MX/WP.21, Geneva: UN, available: http://www.opbw.org/ [viewed 17 January 2010], p. 4.

19 Furukawa 2009, op. cit.

If there has been little evidence of such instruction, it is worth investigating why these subjects have not been incorporated and how faculty members view the relevance of issues such as dual use, biosecurity and biosafety. It is important to identify what kinds of obstacles exist in order to implement such education. In other words, a focused investigation would be able to identify how faculty members recognise 'uncertainties, unknowns, and doubts' about education on dual-use issues, as illustrated in the introductory chapter of this book. By identifying these obstacles it will be possible to envisage effective future polices to help mitigate the current lack of awareness about dual-use issues.

Survey on Biosecurity Education in Japan

In this context, the NDMC in Japan and the University of Bradford in the UK, conducted a survey to analyse the current state of biosecurity education in Japan.[20] The investigation looked at 197 life-science degree courses consisting of 98 undergraduate and 99 postgraduate curricula at 62 universities from 36 different prefectures/regions in Japan. Employing the same basic structure and methodology as the survey on biosecurity education in European universities,[21]the study consisted of two data-collection stages. The first was an online investigation focusing on publicly available syllabi and other information from the websites of life-science degree courses. Specifically, this investigation looked for six possible indicators of biosecurity-education topics. The first three indicators were used to identify the 'presence of modules' on respective subjects within the existing curricula. Thus, the survey asked whether there was evidence of specific modules on 'biosecurity', 'biosafety'[22] and 'bioethics'.

The remaining three indicators were used to identify the 'presence of references' to respective topics within existing modules, even though there are no particular modules on such topics. Thus, the survey asked whether there was evidence of specific references to the following topics within current curricula: dual-use issues; international arms-control or disarmament regimes; and ethical guidelines as well as codes of conduct. The second stage was a follow-up questionnaire to clarify the findings of the online investigation.

---

20 Minehata and Shinomiya 2009, op. cit.

21 Revill 2008, op. cit.

22 In Japan the terms 'biosecurity' and 'biosafety' are used differently. Regarding the former, see note 5. Biosafety measures have been taken in laboratories by safely managing pathogens and toxins with a view to preventing accidental release of bioagents into the field and the exposure of people. Moreover, scientific research on genetic engineering has been taking place internationally based on the Cartagena Protocol on Biosafety to the Convention on Biological Diversity adopted in 2000. Japan introduced the Law Concerning the Conservation and Sustainable Use of Biological Diversity through Regulations on the Use of Living Modified Organisms, which came into force in 2004.

The available information was organised into four categories, as follows:

> *Exist*: refers to data where we can say with a degree of certainty that the required information was present
>
> *Not Exist*: refers to data where we can say with a degree of certainty that the required information was not present
>
> *Unclear*: refers to data where there is some information available but we cannot say with certainty whether the required information exists or not
>
> *Not Available*: refers to data where there are significant constraints upon access to the required information.

Survey Results

Figure 1 shows that the survey identified three specific biosecurity modules and some other instances of biosecurity-specific teaching. Although there were only 18 cases of biosafety modules, biosafety education has been provided in many universities by means other than a single educational component. Bioethics modules were the most commonly found topic in this survey, with 138 examples. In a small number of cases these also dealt with dual-use issues. Some 34 universities included topics of relevance to dual-use issues without using this specific term. References to international prohibition regimes against biological and toxin weapons were also limited, with only 11 cases found. Finally, references to ethical guides or codes were fairly prevalent, with 94 cases largely included in bioethics modules. However, only a small number of ethical guides or codes addressed dual-use issues.

The survey questionnaire asked whether faculty members were familiar with the investigation topics. If they were not familiar with the terms of enquiry, it is very difficult to expect the presence of either modules or references. In view of this, the beginning of the questionnaire asked: 'Have you ever heard about the terms 'biosecurity', 'biosafety' and 'dual use'?' Figure 2 shows the extent of familiarity with these specific references as demonstrated by the questionnaire results. The terms 'biosecurity' and 'biosafety' were relatively well known to the respondents (with 21 positive recognitions). Although this survey broadly defined 'dual use' as referring to the possibility of a misuse of science for hostile purposes, in Japan the term more commonly refers to the possibility of military technology being applied for civilian purposes, and *vice versa*.[23] Thus, the familiarity of respondents with the term within this survey could be expected to be low. Indeed, 17 respondents were not familiar with the reference.

23 Yamada, N. 2008, 'Advances in science and dual-use', presented to the *Symposium on bioterrorism prevention and biosecurity education*, 17 April, Tokyo, Japan.

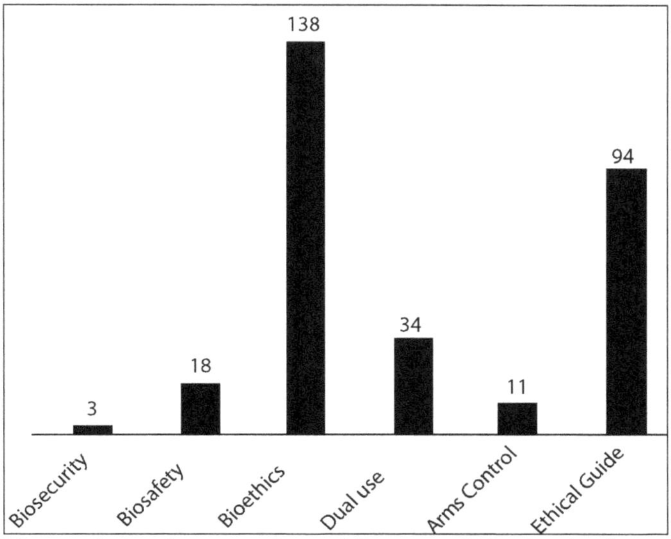

Figure 1: Implementation of Surveyed Topics in Japan

Specific modules on biosecurity, biosafety and reference to arms control have been less developed (n=3, n=17, n=10), whilst modules on bioethics and references to ethical guides were numerous (n=137, n=94).

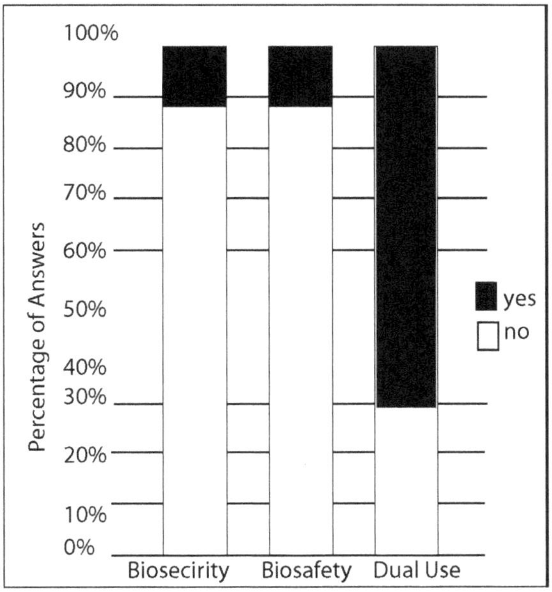

Figure 2: Level of Familiarity with the Terms

The majority were familiar with the terms biosecurity and biosafety (white, n=21), but not with dual use (black, n=17).

Biosecurity Modules

Quantitatively, three cases of biosecurity module were discovered, and all of these were at postgraduate level (see Figure 3). These existing modules were primarily focused on public-health preparedness against the threat of the deliberate use of pathogens or accidental release of diseases. Thus, one module provides an educational course that considers biological risks *vis-à-vis* international public-health policy, mainly focusing on public-health responses to biological and chemical weapons; surveillance of infectious diseases in Japan; and issues related to international public-health policy and processes. Similarly, another module introduces risk-management policy in the public-health sphere in relation to biological and chemical terrorism.

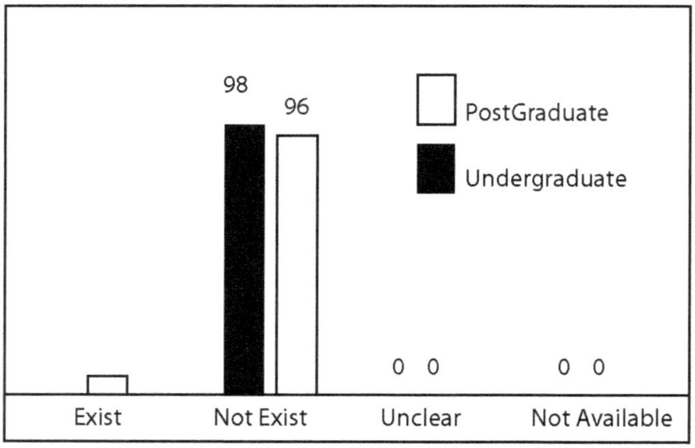

Figure 3: Number of Biosecurity Modules

Only three specific modules on biosecurity have been developed (n=3) at postgraduate level.

Among the universities that had not conducted any biosecurity modules, some have been implementing education in this area on a more *ad hoc* basis by organising seminars/conferences or providing online educational facilities. There is a trend for such events and materials to focus on the following aspects: laboratory biosecurity and biosafety measures, national and local responses to bioterrorism, and emerging and re-emerging diseases. Amongst the examples identified is a medical department which has a 'Bio-Preparedness Wiki' providing an online and open information-exchange platform for users to both download and upload information to this website.[24]

24 Keio University 2009, *Keio Bio-Preparedness Wiki*, available: http://biopreparedness.jp/index.php?MEXTPJ_en [viewed 17 January 2010].

Biosafety Modules

In respect to the introduction of the national legislation to implement the Cartagena Protocol on Biosafety in 2004, universities have set up committees to inform university members and students about the characteristics and physical management of pathogens and relevant national regulations for the prevention of the spread of diseases. Some committees have both laboratory biosafety and biosecurity measures, and mandate their university to provide certain types of education for students. Moreover, there were already some biosafety measures that had been developed in Japan by the National Institute of Infectious Diseases (NIID) and encoded in the 'Biosafety in Microbiological and Biomedical Laboratories' guidelines.[25]

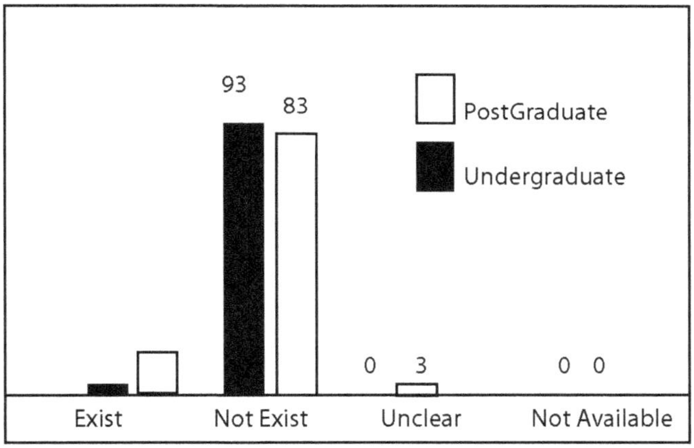

Figure 4: Number of Biosafety Modules

A small number of specific biosafety modules have been developed at undergraduate (black, n=5) and postgraduate levels (white, n=13).

However, as Figure 4 indicates, this does not mean universities are obliged to provide a specific module on biosafety for the purpose of educating students. Nonetheless, 18 cases of particular modules were identified, and more than two-thirds were at postgraduate level. Although we found examples of modules using the term 'biosafety' in subject titles, more generally biosafety measures tend to be mentioned as part of other modules. Some of the biosafety processes are provided in relation to laboratory biosecurity systems to physically contain dangerous pathogens and toxins.

25 Japan 2005b, op. cit.p. 2.

Bioethics Modules

Bioethics modules were found in 138 out of 197 courses in this survey (see Figure 5). Amongst other things, the main components included the history of medicine, the self-determination of patients, informed consent, transplants, gene therapy and counselling. These elements could overlap with ethical guidelines and codes of conduct for scientists such as the Declaration of Geneva of 1948, the Hippocratic Oath and the Declaration of Helsinki of 1964 for medical professionals.[26]

Although the majority of bioethics modules were not framed in the context of dual-use issues, there were some that arguably considered them. For example, there was a module, *Ethics in Human Experiments*, which reviews the history of German, US, and Japanese human experiments since World War I and includes the Japanese bioweapons programme. Another module, *Science Technology and Society,* considers the conduct and dual-use ethics of scientists using nuclear science and nuclear weapons as examples.

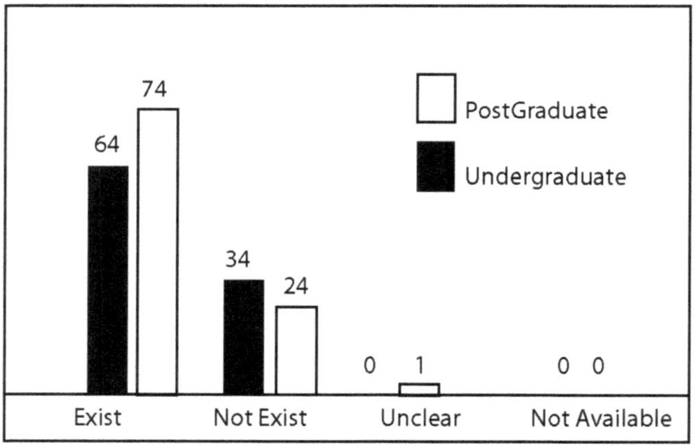

Figure 5: Number of Bioethics Modules Bioethics modules were prevalent at both undergraduate (black, n = 64) and postgraduate levels (white, n = 74).

A graduate school of science gives specific educational content on dual-use issues in the life sciences. A module, *Introduction to Research Ethics*, provides a specific lecture on *Social Responsibility of Scientists: From a Perspective of National Security*, including the issue of dual use and biosecurity. The school also provides a bioethics module, *Bioethical Science*, which considers security and social dangers derived from unpredictable risks in new life-science research. The research centre associated with the school offers seminars including

26 *World Medical Association Declaration of Helsinki: Ethical principle for medical research involving human subjects*,available: http://www.wma.net/en/30publications/10policies/b3/index.html [viewed 17 January 2010].

Promoting Research Ethics: From a Web to Practice in Preventing the Destructive Application of Science. This exhibits one of the most comprehensive approaches towards dual-use issues by indicating the necessity of multifaceted measures to deal with threats posed by the misuse of the life sciences, including biosecurity and biosafety measures.[27]

Dual-Use References

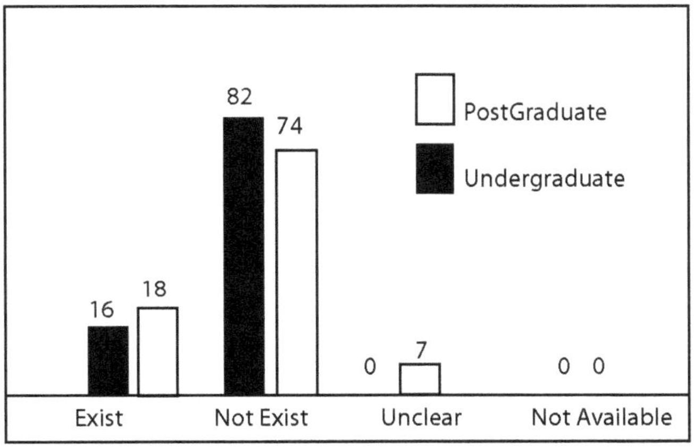

Figure 6: Number of Dual-Use References

The presence of educational content on dual-use issues was notable at undergraduate (black, n=16) and postgraduate levels (white, n=18).

Figure 2 showed the low level of familiarity of faculty members with the term 'dual use'. However, the quantitative results indicate that some 34 universities, of which 16 cases were in undergraduate and 18 in postgraduate courses (see Figure 6), have been providing dual-use content in their academic modules — but, interestingly, without using the term itself; that is, the relationship between science and its potential misuse was a relatively common topic in existing course content at the universities surveyed.

A trend in dual-use content in existing education is illustrated by the history of science and its exploitation for violent purposes in wider fields other than the life sciences. One university had a module on *Science and Society* that considered what science introduced into society, including both social benefits and harmful consequences. The graduate school of the same university also had a module, *Scientific Technology and Society*, which reviewed the historical evolution of science and its dual-use aspects, and included illustrations of chemical and nuclear weapons. A *History of Science* module in another university demonstrated the

27 The second ASMeW international ethics seminar, available: http://www.waseda.jp/scoe/sympo/080204seminar/080204e.html [viewed 17 January 2010].

scientific evolution from antiquity to the modern times of mass production and mass destruction. There was a further example, more specifically, *Introduction to Medical Zoology*, which studied a diverse range of animals and insects. Alongside its research on toxins for pharmaceutical purposes, it discussed considerations such as the many types of biotoxins that have been developed for biological weapons that can also be of concern in biocrimes.

Arms-Control References

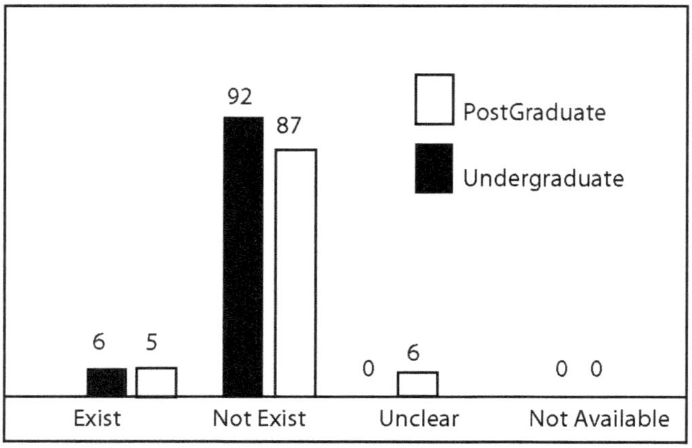

Figure 7: Number of Arms-Control References

This was the second least prevalent topic in this investigation at undergraduate (black, n=6) and postgraduate levels (white, n=5).

This topic was one of the most unfamiliar themes of education found in this survey, with only 11 cases in total (see Figure 7). Unless faculty members had a specific interest in security issues, international prohibition regimes against biological and chemical weapons were not included as part of science and medical teaching. However, an undergraduate course had a forum containing a series of online papers,[28] one of which provided a brief illustration of modern biowarfare programmes worldwide, with a specific focus on smallpox and anthrax. Also, the forum highlighted the potential threat of bioterrorism, using the case of Aum Shinrikyo and the anthrax attacks after 11 September in the US, to illustrate contemporary concerns over this risk. Having considered those dangers, the series moved on to cover the international prohibition against biological weapons, including the 1925 Geneva Protocol and the 1972 Biological and Toxin Weapons Convention, and what should be done in Japan to strengthen these regimes. Another module at a faculty of science entitled

28 See Research Center for Animal Life Science Shiga University of Medical Science 2008, Primate forum: Lectures on Zoonotic diseases, available: http://www.shiga-med.ac.jp/~hqanimal/ [viewed 17 January 2010].

Chemistry, referred to the Chemical Weapons Convention (CWC), whilst the Department of Chemistry listed the domestic laws undertaken to implement the CWC within the Department.

Ethical Guidelines

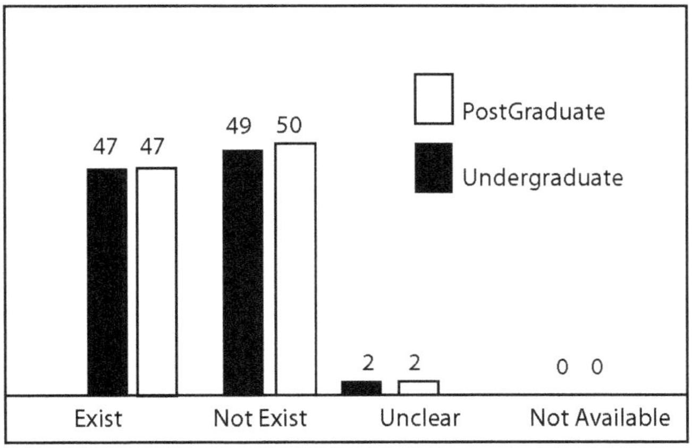

Figure 8: Number of Ethical Guideline References

Usually, being referred to within bioethics modules, this topic was widely provided at undergraduate (black, n=47) and postgraduate levels (white, n=47).

Some universities have been implementing education in this area not as full modules but with reference to relevant ethical guidelines within courses on other topics. As Figure 8 shows the references to these standards or codes of conduct for scientists have a relatively high degree of presence in our survey with 47 cases both at undergraduate and postgraduate levels.[29] In addition, investigation results from the more prevalent bioethics modules could indicate a much higher presence of references to ethical guidelines or codes, because of the close overlap between both topics.

These topics are primarily provided with a view to ensuring good practice in medicine or preventing misconduct in scientific research rather than promoting understanding of dual-use issues. Most of the guidelines that could be found were in relation to research areas on the human genome, genetic engineering and human embryonic stem (ES) cells. As was the case with Living Modified Organisms (LMO) and biosafety, these have been guided by respective government regulations. However, some dual-use references could be recognised. A school of medicine noted that their university did not authorise a patent to a product of scientific research if it raised concerns regarding public safety or weapons development. Another university listed the websites of the

29 For further information, see Hara, S. and Masuda, K. 2007, 'Current state of institutional review boards (IRB) of special functioning hospitals in Japan', *Clinical Evaluation*, vol. 35(2), pp. 375–408.

World Health Organisation (WHO) and the US Center for Disease Control and Prevention (CDC) with a view to providing information on biological, toxin and chemical weapons, and also references to incidents of bioterrorism with anthrax.

Existing Interest in Biosecurity Education

The survey results indicated that there was a clear lack of educational topics on biosecurity despite a certain level of presence on dual-use references. However, this does not necessarily mean there is a lack of interest in such education. In cases where there was no such module or reference at the investigated university, the questionnaire provided multiple-choice options related to each topic: 'Although we (the investigated university) have not provided such a topic: A, we should implement the topic; B, we are interested in the topic but it is difficult to implement at the own university; C, we do not think it is necessary for our academic curricula'.

Figures 9 (undergraduate level) and 10 (postgraduate level) suggest that many universities had a positive interest in the subjects in general, especially the implementation of research guides or codes of conduct at postgraduate level. Nevertheless, the majority of the feedback also noted that it was difficult to introduce such topics in their current academic environment. The respondents also suggested there were a series of difficulties that caused the lack of provision for such education. These are elaborated in the following section.

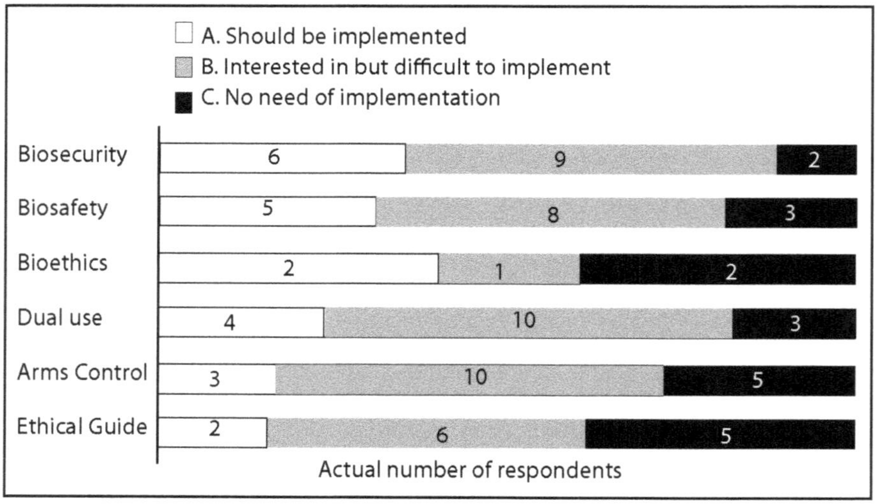

Figure 9: Interests of Faculties on Educational Topics: Undergraduate

Except for bioethics modules (gray, n=1), the majority of all other educational topics were categorised into 'B. Interested in but difficult to implement'. The interest in biosecurity modules (gray, n=9), dual-use issues (gray, n=10) and arms control (gray, n=10) was high. Numbers on 'C. No need for implementation' were also notable for arms control (black, n=5) and ethical guidelines (black, n=5).

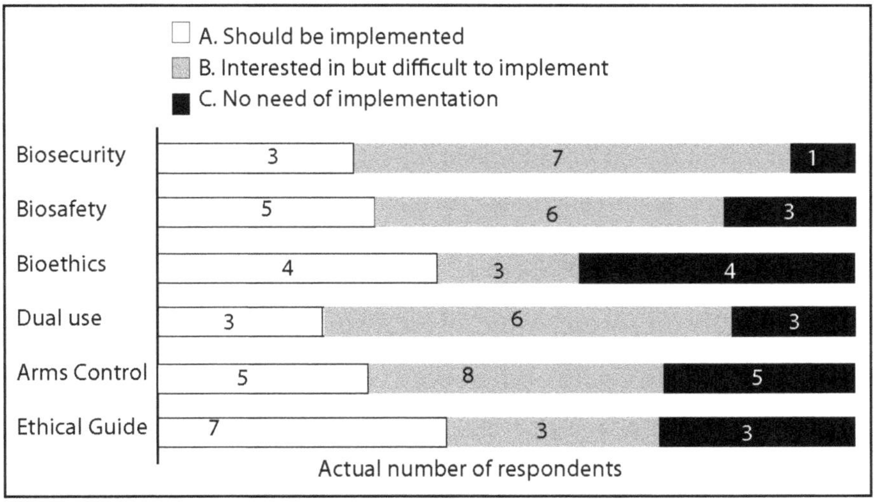

Figure 10: Interests of Faculties on Educational Topics: Postgraduate

At postgraduate level, references to ethical guidelines for research scored good interest with faculty members (white, n=7), a contrast to undergraduate level (see Figure 9). There was relatively high interest on biosecurity modules (gray, n=9), dual-use issues (gray, n=6) and arms control (gray, n=8), but the majority of answers indicated that it is difficult to implement within the current curricula.

Obstacles to the Implementation of Biosecurity Education

Although the limited number of responses from 48 departments in 24 universities does not permit a statistically significant generalised analysis, it does illustrate the difficulties faced by university lecturers. We found that the responses to our questionnaire indicated there was:

- an absence of space in the existing curricula
- an absence of time and resources to develop new curricula
- an absence of expertise and available literature on biosecurity education
- doubt about the need for biosecurity education.

Dual-Use Bioethics: A Starting Point

Despite the commonly addressed obstacles, the survey still indicates the possibility of promoting biosecurity education in Japan. Existing curricula, including courses on bioethics, would be a key intervention point for its introduction. By integrating biosecurity considerations into existing bioethics education, the ethical considerations of life scientists can be expanded to include the potential misuse of scientific knowledge. 'Dual-use bioethics' is one possible starting point to develop biosecurity education in the current Japanese educational environment.

Dual-Use Bioethics: An Additional Questionnaire

In order to research the possibility of dual-use bioethics education, the survey team conducted an additional questionnaire by specifically targeting lecturers on bioethics education at some of the same universities. The new questionnaire asked, 'What kind of obstacles can be expected in the process to introduce dual-use issues into the existing ethical education for life scientists?' As in the earlier survey, some of the respondents identified similar uncertainties, unknowns, false starts, and doubts towards biosecurity education.

However, at the same time they addressed a certain possibility of dual-use bioethics education. One respondent noted, 'Even without a specific module on dual-use issues, introductory education for bioethics can cover dual-use issues'. Another pointed out in this regard that within the already busy curricula of universities it was of critical importance to achieve 'understanding [of the topic] and coordination amongst relevant stakeholders in the department, particularly by personnel from each research division, to develop academic administration'. To indicate how awareness of matters pertaining to dual-use bioethics could be introduced, another respondent commented that he felt that is was worthwhile and that he would 'cover the issue of dual use, including the case of Aum Shinrikyo, for about 15 minutes' in his lecture at the faculty of life science, from the following semester.

In addition to such bottom-up approaches, some top-down methods also prove useful in promoting dual-use awareness. In view of this, the additional questionnaire also asked whether ethical awareness of dual-use issues among life scientists should be used as an assessment criterion for grant applications to funding bodies, for the review processes of scientific journals, or for ranking systems of universities.

The responses to these questions varied. For example, on a positive note, one contributor noted that 'many Japanese scientists are internationally recognised with their scientific research through publication in top scientific journals, but they should also make additional efforts in order to assure the international confidence in ethical awareness of scientists about dual-use issues...and those top-down methods facilitate the latter'. Another respondent argued that 'although the establishment of such assessment criteria is necessary, raising awareness [among relevant actors involved in this evaluation process] should be a higher priority'.

This point was reinforced by other respondents, who pointed out that it would be preferable to assess the level of awareness of scientists of ethical conduct in dual-use issues and their practice to prevent the potential misuse of science. The question regarding whether awareness of such issues should be used in ranking universities, however, would require careful consideration. As one respondent noted, different universities 'put different emphases on science education. Some do not have faculties of science or medicine but are teaching bioethics in the social science faculties.' Whether the same standard of ethical awareness should be required within social science and natural science faculties remains a matter for further discussion.

Need for an Accessible Educational Resource

As Malcolm Dando argues elsewhere in this volume, placing open-source teaching material online, via the internet, could assist in the assimilation of biosecurity education into existing curricula, ease constraints on time spent planning and preparing material, overcome financial constraints on the development of biosecurity programmes, and provide the expertise required for efficient and effective integration of such material.[30]

One of the Japanese lecturers who responded to the questionnaire commented that 'comprehensive educational material is welcome, but what may be more useful would be a concise scenario-based education, backed up by audio-visual material to catch the attention of students'. The lecturer explained this is because 'many university lecturers and students in science departments may feel distant from the dual-use topics'.

By using such educational resources, a next stage to promote biosecurity teaching would be to build capacity through the implementation of such modules in different academic contexts and institutions. Knowledge gained during this process could then be used to develop best-practice standards. Specifically with regard to the second and the third stages, the following sections illustrate the experience of the NDMC.

Biosecurity Education at the NDMC in 2008 and 2009

By using a freely available online teaching resource, specifically designed for facilitating easier implementation of biosecurity education for university lecturers (that is, for 'train-the-trainers programmes'[31]) biosecurity educational agendas were provided at the NDMC in October 2008 and March 2009.[32] The

30 R. 2008, 'Developing educational modules for life scientists accelerating the process though an open source initiative', presented to the *IWG–LNCV Biological workshop and round table on fostering the biosecurity norm: An educational module for life sciences students*, 27 October at the Municipality of Como, Italy.

31 University of Bradford 2009, *Dual-use Bioethics.net*, available: http://www.dual-usebioethics.net/ [viewed 17 January, 2010].

32 Minehata, M., Yamada, N., Kobayashi, Y., Shinomiya, N., Miyahira, Y., Dando, M. R. and Whitby, S. M. 2009, 'Developing an Educational Module Resource for Life Sciences through the Biological and Toxin Weapons Convention', paper presented to the *2nd Biosecurity symposium*, 9–10 February Sydney, Australia.

teaching programme in 2008 involved a five-day course for 19 postgraduate students at the beginning of their graduate degree in Medicine (see Table 2). This process was carried out using similar content for 57 medical students at the end of their six-year curriculum in 2009. These topics have now been integrated more systematically into the syllabus to raise the awareness of students.

Table 2: Outline of Dual-use Education at the NDMC Graduate Course of Medicine[33]

Day	Time	Topic
Day 1	09:00-09:15	Introduction
	09:15-10:45	Life Science and Ethics
	11:00-12:00	Intellectual Property
Day 2	09:00-09:45	Codes of Conduct for Life Scientists
	09:45-10:30	Dual-use Dilemma: History and Outline
	10:45-11:30	Biological and Toxin Weapons Convention*
	11:30-12:00	Present Status of Biosafety
Day 3	09:00-09:50	Biosecurity: Research Field of Concern
	09:50-10:40	Surrounding Situation of Scientists and Scientific Papers
	10:50-12:00	Ethics for Animal Experiments: Basic Rules and Legislations
Day 4	09:00-10:00	How to Search Scientific Papers
	10:10-11:20	How to Use Statistical Analysis System (SAS)
	11:20-11:50	Examination
	13:00-14:30	Guidance of Core Facilities
Day 5	09:00-10:00	Feedback and Discussion 1
	10:10-11:20	Feedback and Discussion 2
	11:20-11:50	Closing Remarks

Note: * NDMC staff modified the online educational module resource to tailor the content into the scheduled educational time available for the course. For the online educational module resource, see University of Bradford 2009, Dual-use bioethics.net, available: http://www.dual-usebioethics.net/ [viewed 17 January 2010].

The students' understanding of the course content was tested using multiple-choice questions. The open-ended questions presented in the feedback and discussion sessions enabled students to give their views on such areas as whether they thought their own research could be a cause for concern. While some acknowledged that this could be the case, others pointed out that the awareness of scientists plays an important role in preventing possible threats, since it is difficult to verify malicious intent in life-science research. At the same

33 Source: Shinomiya, N. 2008, 'Developing the material required for mandatory dual-use education of life scientists (Part 2)', presented to the *IWG – LNCV Biological workshop and round table on 'Fostering the biosecurity norm: An educational module for life sciences students'*, 27 October at the Municipality of Como, Italy.

time, some argued that regulations introduced without careful consideration might harm the scientific freedom of individuals without producing effective results.

In discussing the kinds of research in the life sciences that could give rise to concern, many students referred to the Fink Committee's report on the seven categories of research areas of concern.[34] Some students paid specific attention to synthetic biology. A recurring response was that synthetic biology could enable researchers to manipulate a virus's antigenicity, pathogenicity and toxicity to a great extent. Misuse of such agents makes effective prevention and response difficult. In relation to biosecurity measures, one student pointed out that 'without a host or parent virus, new viruses can be cultured based on DNA sequences and chemical synthesis...therefore, misuse cannot be prevented solely by physical control of biological agents' and would require greater governance of information and expertise.

Finally, an anonymous questionnaire was circulated to check the accessibility of each taught topic in this educational process by asking whether 'your understanding on the following aspects of the module was developed' Scoring five indicated the highest positive mark and one, the lowest. The results of the questionnaire, shown in Diagram 1, indicate that students assessed the programme very positively.

There are several lessons that can be learned from this process. Firstly, since the NDMC programmes were specifically designed for the medical students at a defence college, these programmes may not necessarily prove useful for other universities in Japan. Secondly, in light of ever-advancing life-science research, the educational content needs to be constantly updated to ensure that scientific and technological discussions about dual-use issues remain up-to-date and relevant. Furthermore, a clearer assessment framework to measure the impact of biosecurity education needs to be developed to show the value of such teaching as an academic subject.

On the positive side, the example at the NDMC demonstrates how an open-source educational module resource can be modified for specific teaching purposes in busy curricula at the university. It also shows how raising awareness is possible through lectures on essential regulations in biosecurity and letting students consider the potential dual-use consequences of their own research. It is recommended that the sharing of information among universities on experiences and lessons learned in this field should be further promoted to develop best practices in biosecurity teaching.

34 Including research on making virus resistant to a vaccine, enhancing the virulence of a pathogen or modifying the host range of a pathogen. See National Research Council 2004, *Biotechnology research in an age of terrorism*, Washington, DC: National Academies Press.

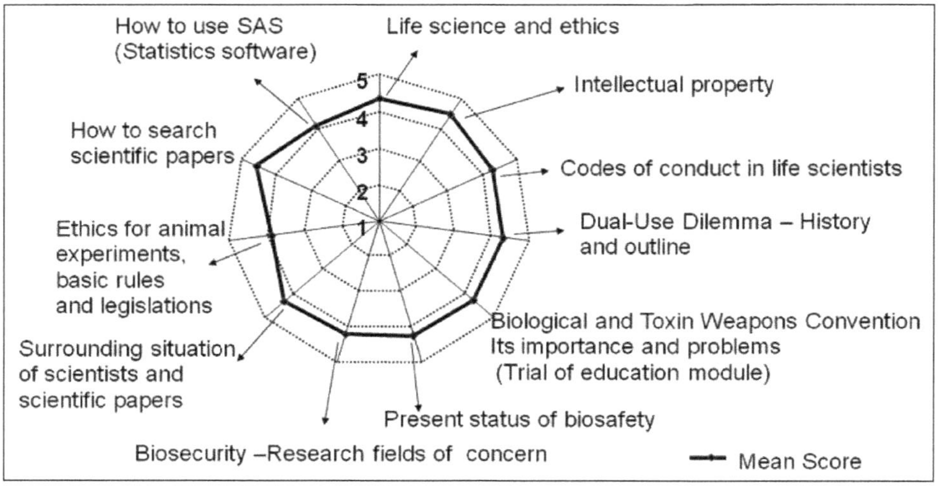

Diagram 1. Mean Score of Questionnaire by the Students on the NDMC Module[35]

Conclusion

The experience of Japan indicates that university-level ethical education on dual-use issues for life scientists can be implemented successfully. The key stages of the strategy adapted in Japan included:

- surveys of existing educational courses
- contact with and among lecturers
- setting up national and international networks
- provision of assistance
- resurveys to check the implementation of education.

A focused survey is useful for investigating the current state of biosecurity teaching or educational content on dual-use issues for life scientists at institutions for higher education. At the surveyed universities, biosecurity modules, followed by arms-control and dual-use references, were the least prevalent topics within the life-science degree courses. However, universities with no biosecurity in their curricula expressed an interest in its introduction. The survey also helped reveal possible reasons why such content had been missing from existing curricula. The identification of 'uncertainties, unknowns, false starts, and doubts' can be an essential understanding prior to an effective

35 Shinomiya 2008, op. cit.

policy-making process to help remove existing obstacles. The survey recognised a possible starting point of biosecurity education by integrating dual-use content into existing bioethics teaching.

A second conclusion to be drawn is that developing contact with (and among) university lecturers is important. This enables universities to share their experience of the implementation of biosecurity education to pursue best practice in such teaching. Importantly, commonly addressed obstacles by university lecturers to introduce such education seem to be structurally embedded and should be dealt with by efforts not only from individual universities, but also from government experts on security and education issues.

Thirdly, for this purpose, setting up national and international networks will be important. Through this process, the promotion of biosecurity education can accommodate the interests of practising scientists and security policymakers and make it possible to strike an appropriate balance between the freedom of scientific research and oversight of science for national security requirements. Education within scientific communities, such as domestic scientific associations and international research groups, is also a key factor to raise biosecurity awareness among life scientists.

The Research Institute of Science and Technology for Society (RISTEX) of the Japan Science and Technology Agency is an important initiative in setting up a national network of 'a few hundred stakeholders in biosecurity, including officials of all relevant ministries and agencies, and experts of universities and research institutions as well as journalists'.[36] This platform will play an essential role to help promote biosecurity education in Japan. Another project is an international seminar framework and a 'Safety and Secure Science & Technology' project supported by the Ministry of Education, Culture, Sports, Science and Technology (MEXT), through which rules for preventing the malign use of science, the introduction of biosecurity/dual-use bioethics training programmes, strategies for developing secure research environment, and so on, can be discussed.

In 2006, the Science Council of Japan (SCJ) introduced a code of conduct for scientists, partly as a basis for an assessment criterion for grant applications to the Council.[37] Though the code was not originally developed to promote ethical responsibility to prevent potentially dangerous consequences of dual-use life science research,[38] Katsuhisa Furukawa argues that the SCJ code was 'drafted in such a way as to cover dual-use risks' as it underlines the ethical responsibility

36 Furukawa 2009, op. cit.

37 Science Council of Japan 2006, *Statement: Code of conduct for scientists*, available: http://www.scj.go.jp/ja/info/kohyo/pdf/kohyo-20-s3e-1.pdf [viewed 17 January 2010].

38 Ibid.

for the safety and security of society as well as educational programmes at research institutions.[39] However, as Furukawa points out, the SCJ code has not been extended to explicitly deal with dual-use issues.[40] The problem is that, to date, the academic community in Japan has paid little attention to this point and no associations for medical or life sciences are taking positive action to prevent the malign use of scientific knowledge. No Japanese scientific journals have so far introduced biosecurity review systems. Providing opportunities for life scientists to learn dual-use examples through congress seminars may solve these problems.

The provision of assistance may also include an accessible and shared education resource. The benefits of developing an open-source biosecurity education programme were demonstrated by the NDMC experience. Other possible provisions may include making ethical awareness of dual-use issues an evaluation criterion for grant applications by funding bodies or review processes for scientific journals, as well as a ranking system for universities. However, such processes require further discussion. A re-survey to analyse the implementation of education will be necessary if further provisions of assistance are provided from scientific, academic and government bodies, to facilitate the accumulation of several examples of biosecurity teaching in Japan. This process will enable the evaluation of such programmes and the sharing of knowledge of best practices of biosecurity education in Japan.

Finally, it should be noted that the series of strategic elements to implement biosecurity education in Japan would not necessarily be sufficient or available in other countries. However, Japan's experience may prove valuable to others nevertheless, whilst Japan also has much to learn from other nations. Indeed, the significance of raising awareness among life scientists should be recognised, at least amongst the member-states of the Inter-Academy Panel (IAP).[41] To achieve securer advancement of the life sciences in the twenty-first century, implementation of biosecurity education and a sharing of knowledge need to be coordinated nationally and internationally, backed up by the active engagement of both scientists and other social actors.

---

39 See the section of codes of conduct in Furukawa 2009, op. cit.

40 Ibid.

41 Inter-Academy Panel 2005, op. cit.

Chapter 6: Bioethics and Biosecurity Education in China: Rise of a Scientific Superpower

MICHAEL BARR AND JOY YUEYUE ZHANG

This chapter explores ethics, education and the life sciences in China. It is based on work conducted by the authors in two separate but complimentary projects. [1] Barr's observations derive from interviews and discussions in Beijing, Shanghai and Guangzhou with life scientists and policymakers in infectious-disease hospitals, university-research labs, the Chinese Academy of Sciences, and the Ministry of Health. Zhang's study focused on China's governance of stem-cell research and involved interviews with scientists, ethicists and policymakers at more than 25 sites across China. Below, we set the context by describing the role of science in China's quest to become a leading power and then consider the place of bioethics within China. We follow this with a discussion of three key areas that have impacted our work and describe some of the lessons we have taken from our experience for future research on bioethics education and biosecurity in China. We conclude with a set of suggestions about what can be done to further biosecurity awareness within China.

For the sake of clarity, we should note our use of terms. By 'biosecurity' we refer to the protection and control of pathogens and toxins to prevent their deliberate theft, misuse, or diversion for the purposes of biological warfare or terrorism. According to our use of the word, this includes researchers' personal knowledge, choices and behaviour, as well as society's collective responsibility to safeguard a population from the dangers of pathogenic microbes. We use the term in contrast to 'biosafety', which we see as laboratory procedures and policies aimed at reducing accidental exposures. Contained within the term 'biosecurity' is the dual-use dilemma, which refers to the possibility that the same scientific research, products, or facilities which are meant for social good could also have an unintended result of threatening a population, either

1 We gratefully acknowledge the support of our funders for this research. The Alfred P. Sloan Foundation in the US and the UK Department of Universities, Innovation and Skills funded Barr's work. Zhang held a Wellcome Trust Biomedical Ethics Studentship Award for her study.

inadvertently or through a deliberate act of bioviolence. Finally, by 'awareness-raising', we accept the Biological and Toxin Weapons Convention's (BTWC) definition of 'improving dialogue and communication' between all 'relevant stakeholders, including policymakers, the scientific community, industry, academia, media and the public in general'.[2] However, our own work (and thus our focus in this chapter) concentrates mainly on educational and informational initiatives aimed at the scientific and academic community in China, mainly life scientists and ethicists.

Rise of a Scientific Superpower

The rise of China constitutes one of the greatest stories of the early twenty-first century. China's impact extends across nearly every sphere — from international finance and trade, to culture and soft power, to global security and weapons proliferation. In the West, when observers discuss Chinese power, there is often a fear palpable beneath the surface: will this country of 1.3 billion people seek to play by the rules of the game (that is, be 'status-quo') or will it become a revisionist state, intent on re-writing the norms of the international system in line with its own perceived interests?[3] And on this question, perhaps no sector raises more concern than the nexus between security and technology.

Since the 1990s science has been a cornerstone of Chinese development. Expenditure on research and development rose from 0.6 per cent of GDP in 1996 to 1.4 per cent in 2006.[4] During this time, the Chinese government set up numerous policies to attract talented life-science researchers, including 'the one thousand talents' campaign which seeks to lure foreign-trained Chinese scientists back home with the contractual promise of high salaries and state-of-the-art facilities.

Beijing's ambitions to become a world leader in science and technology should not be underestimated. President Hu Jintao noted in 2006 that China's general goal was to 'visibly increase the country's indigenous innovation capacity, visibly increase the capacity of science and technology to promote economic and social development and guarantee national security, visibly increase the overall research strength of basic sciences and frontier technologies, strive for scientific and technological achievements of major world implications'.[5]

2 Biological and Toxin Weapons Convention 2007, Report of the 2007 Meeting of States Parties, BWC/MSP/2007/5, para. 21.
3 Johnston, A. 2003, 'Is China a status quo power?' *International Security*, vol. 27, pp. 5–56.
4 Chen, Z. 2008, 'Biomedical science and technology in China', *The Lancet*, vol. 372, pp. 1441–3.
5 Zhu, Z. and Xu, G. 2008, 'Basic research: Its impact on China's future', *Technology in Society*, vol. 30, p. 296.

These comments were soon supported by very real and quantifiable goals:

> By the year 2020, the ratio of gross expenditure on research and development will rise to over 2.5 per cent; the contribution rate of scientific and technological advance to economic growth will rise to over 60 per cent; the dependence on foreign technologies will drop to less than 30 per cent; and the annual granting of invention patents to Chinese nationals and the international citation of scientific papers will rank among the world's top five.[6]

Approximately 20 per cent of this total investment goes to the life sciences, an area in which China has made considerable progress.[7] As early as 2000, Beijing-based genomic centres were estimated to have had more sequencing capacity than France and Germany combined.[8] Whilst the most visible success of Chinese life scientists was the 2002 decoding of the rice genome, China has attracted the attention and interest of both European and US audiences across a number of fields.

China's great leap forward in science and technology begs the question whether similar gains have been made in attempts to socially regulate emerging technologies. The answer, as in most countries, is mixed. Some still portray China as a 'Wild East', a place where scientific progress need not be hampered by ethical reservations or public unease.[9] With doubts over standards in regulatory enforcement and of transparency, combined with large amounts of funding for scientists who are under pressure to generate results, and a culture that — on the surface at least — seems to downplay the importance of individual rights, critics contend that biomedical research in China remains ethically problematic.[10] However, what such accounts miss is the sizable body of regulation pertaining to biological research, including guidelines on biomedical studies involving human subjects, specific advice on clinical drug trials and embryonic stem-cell research, as well as provisions for good clinical and laboratory practice. Crucially, informed-consent requirements, provisions for review by ethics committees, and legal sanctions feature prominently in many regulations.[11] Of course, germane to successful regulation is education and awareness-raising.

---

6 Ibid.

7 Chen, Z., Wang, H. G., Wen, H. J. and Wang, Y. 2007, 'Life sciences and biotechnology in China', *Philosophical Transactions of The Royal Society*, vol. 362, pp. 947–57.

8 Schneider, L. 2003, *Biology and Revolution in Twentieth-Century China*, Lanham: Rowman & Littlefield.

9 Wilsdon, J. and Keeley, J. 2007, *China: The next science superpower? The Atlas of Ideas: Mapping the new geography of science*, London: DEMOS, available at: www.demos.co.uk

10 Hennig, W. 2006, 'Bioethics in China', *EMBO Reports*, vol. 7, pp. 850–4.

11 Medical Research Council 2009, *China-UK Research Ethics (CURE) Committee Report*, London: Medical Research Council.

Nascent but Growing: Bioethics Education in China

Medical ethics teaching in China began in the 1980s. However, early courses were limited to abstract (and solely Western) theories and principles of general ethics. In order to help boost student interest in ethics, Chinese lecturers, with assistance from The China Medical Board in New York and the Chinese Society of Medical Ethics (formed in 1988), sought to reform teaching to incorporate ideas and cases relevant to Chinese culture.[12]

A number of factors have since contributed to the steady growth of ethics teaching.[13] First, the jump in numbers of science students in China has meant that universities have had to expand their provision of ethics courses. Postgraduate programmes in biomedical ethics have been established in a number of key institutions, with the first Master's degree being offered at Tianjin Medical University in 2000. A second reason is the manifold problems faced by China's public-healthcare system. The benefits of China's rapid economic growth have been uneven, badly affecting healthcare services. For instance, it is estimated that only 10 per cent of the rural poor have adequate facilities for sanitation, whilst less than 30 per cent have a reliable source of drinking water.[14] Exacerbating these problems is the plight of China's migrant population: 140 million people are excluded from public medical insurance as they move between cities and the provinces in search of improved opportunities.[15] In this context, it is not surprising that China has sought to develop greater capacity to study, debate, and solve such issues that put a strain on the healthcare and biomedical establishment.

The 'Software' Problem: Ethics Education as a Biosecurity Strategy

The connection between public health, public policy, and ethics education is evident when examining biosecurity in China.[16] Chinese leaders were deeply embarrassed by the severe acute respiratory syndrome (SARS) outbreak in 2003, which some estimate cost the country US$10 billion in lost tourism revenues

---

12 Yali, C. 2003, 'Comparison of medical ethics education between China and the United States', in Song, S. Y., Koo, Y. M. and Macer, D. (eds), *Bioethics in Asia in the 21st Century*, New Zealand: Eubios Ethics Institute; Doering, O. 2003, 'Teaching medical ethics in China. Cultural, social and ethical issues', in Song, Koo, and Macer (eds), op. cit.

13 Li, E. C. 2008, 'Bioethics in China', *Bioethics*, vol. 22, pp. 448–54.

14 Tang, S., Meng, Q., Chen, L., Bekedam, H., Evans, T. and Whitehead, M. 2008, 'Tackling the challenges to health equity in China', *The Lancet*, vol. 372, pp. 1493–501.

15 Hu, S., Tang, S., Liu, Y., Zhao, Y., Escobar, M. L. and de Ferranti, D. 2008, 'Reform of how health care is paid for in China: Challenges and Opportunities', *The Lancet*, vol. 372, pp. 1846–53.

16 Barr, M. 2009, 'China's role as a biosecurity actor', in Rappert, B. and Gould, C. (eds), *Biosecurity: Its Origins, Transformations and Practice*, Basingstoke: Palgrave Macmillan.

alone.[17] Whilst fears that the epidemic would spread to rural areas were not realised, SARS nonetheless highlighted the inequalities of China's healthcare system. Its impact on China can hardly be understated. One microbiologist and biosecurity expert at the Chinese Academy of Medical Sciences explained that SARS was as important to China as the 11 September terrorist attacks were to the US in terms of their political, economic and psychological consequences.

Whilst China responded to the crisis with a range of new regulations (including revision of its 1989 Law on the Prevention and Treatment of Infectious Diseases), the need for greater education and training was brought rudely to attention again in 2004. A batch of the SARS virus at the National Institute of Virology in Beijing, mistakenly thought to have been inactivated, was moved from a BSL-3 storage container to a non-regulated lab where medical students were working on diarrheal diseases. The breach of security subsequently resulted in eight infections and one death, as well as the temporary closure of the Institute and quarantine of over 700 individuals suspected of coming into contact with the virus. The accident had clearly been the result of human negligence. One microbiologist at Fudan University refers to this as a 'software' problem — meaning that whilst much attention has been paid to the 'hardware' (the building of hi-tech labs, autoclaves, cabinets, locks, doors, and so on), the human element of biosecurity has been neglected. That is, the training, behaviour, management skills, expert knowledge, and duties of care needed to operate high-level laboratories safely have not kept pace with the introduction of new facilities.

The 'software' problem includes much more than lab safety. Yet statements by the Chinese Delegation to the BTWC Meeting of Experts show that their focus is almost entirely on safety, not the wider issue of dual use. According to their declaration, biosecurity 'education and awareness raising' refer solely to 'laboratory safety management and technical training, biosafety licensing, preparedness for health emergency and response and veterinary biosafety'.[18] These efforts are obviously important and are to be supported. However, like other countries with growing biotechnology sectors, China's adoption of educational measures and codes of conduct aimed at addressing a broader agenda of oversight of the life sciences and how biological research might be exploited for illegitimate purposes, remains uneven. A small number of top universities and scientific associations, including the Chinese Academy of Sciences (CAS) have sought to establish an internal code of ethics that aims to promote scientific ethics, as well as the integrity and moral character of staff. CAS has also set up

17 Wen, H. 2004, 'The Short Term Impact of SARS on the Economy', *Asian Economic Papers*, vol. 3, pp. 57–61.

18 Chinese Delegation to the Biological and Toxin Weapons Convention Meeting of Experts 2008a, Statement made on Biosafety & Biosecurity Capacity Building; Chinese Delegation to the Biological Weapons Convention Meeting of Experts 2008b, Poster on *China's Practice on Biosafety and Biosecurity*.

a special commission for scientific integrity to promote transparency, autonomy and accountability of research. These types of codes are to be encouraged and broadened to specifically promote dual-use awareness. Yet it must be noted that CAS is essentially the scientific arm of the government, supported by the State Council itself, and considered to be the most prestigious scientific institution in the country. Whilst bodies like CAS may set a useful example, the real challenge lies in reaching provincial and district-level labs, especially outside the main urban settings, where it is harder to monitor activities.

It is important to reiterate that the amount of attention paid to dual-use issues varies according to the site. In 2002, China's main legislative body, the State Council, passed two sets of regulations regarding dual-use equipment and technologies.[19] The directives contained measures to strengthen export controls to prevent the diversion of dual-use biological agents, related equipment, and technologies that could be used in weapons production. It also included an export-licensing system and provisions for the criminal prosecution of domestically based violators. Significantly, the export-control list covered within the regulations provided an extensive list of pathogens and toxins, thus putting China in accord with control lists of the Australia Group (to which it still does not formally belong). Whilst we have no reason to doubt the enforcement of these regulations, on the wider issues of personal responsibility, we found that most Chinese life scientists we interviewed were not particularly concerned about the dual-use implications of their work and did not regard bioterrorism or bioweapons as substantial threats. The reasons for this varied, but, as in the West, many scientists in China tend to view scientific progress as inevitable and generally think that pressures to publish and present findings mean that research will, one way or another, be conducted and find its way into the public domain.[20] Our findings were confirmed by the views of participants at China's first ever, international meeting dedicated to the dual-use dilemma. Organised by CAS in 2008, with the support of the Inter Academy Panel and the Organisation for Economic Co-operation and Development (OECD), delegates to the workshop expressed concern over dual-use issues but also agreed that different stakeholders tended to view the concerns differently. Compared to security specialists, scientists in general were seen as being unaware of the potential dual-use nature of their research.

The remainder of this chapter will address several themes from our experience of researching and promoting bioethics in China and propose a series of practical steps that could help promote biosecurity education.

---

19 State Council of China 2002a, *Regulation of the People's Republic of China on Export Control of Dual-Use Biological Agents and Related Equipment and Technologies*; State Council of China 2002b, *Dual-Use Biological Agents and Related Equipment and Technologies Export Control List*.

20 For more on scientists' levels of awareness, see Rappert, B. 2007, *Biotechnology, Security, and the Search for Limits*, Basingstoke: Palgrave Macmillan.

Lessons Along the Science/Society/Security Interface

Ethicisation: How an Issue Becomes an Ethical Problem

In the West, social scientists working in the area of science and security studies are alert to how key actors transform certain issues into security concerns. This process is referred to as 'securitisation' and involves studying 'who securitises, on what issues, for whom, why, with what results, and under what conditions'.[21] Similarly, during our work in China we came to see that it was necessary to heed attention to the way in which an issue becomes a source of bioethical (and biosecurity) concern. Zhang has coined the term 'ethicisation' to refer to a process of raising awareness.[22] Its function is to map out the social, legal, political and financial concerns that scientists should take into consideration. Like securitisation, ethicisation embraces no specific moral objectives. It can, however, lead to a re-evaluation of existing judgments.

In this context it is important to remember that life-science research cannot be conducted without international collaboration. Chinese scientists are not only heavily involved in joint projects with Western colleagues but participate in a multitude of exchange programmes and seek to publish their results in English-language journals. So, as transnational investment and communication become standard practices, 'ethical concerns' in need of regulation become 'infectious'. The necessity to facilitate cooperation has resulted in the request for increased compatibility of local frameworks with that of the potential partners. Consequently, bioethics is no longer a segmented social aspect rooted solely in a specific cultural milieu. The perception of research priorities is shaped as much by national factors as they are by debates in the global scientific and bioethical communities. And this, of course, has knock-on effects for how ethics is taught.

One example may help to shed light on this process. On a visit to a regional headquarters of the China Hematopoietic Stem Cell Data Bank, the director told Zhang how visiting Western stem-cell banks expanded her awareness of ethical issues. In the original planning of their cell bank's office space, people walk in and are first welcomed by a whole wall of glass-panelled covered shelves, with hundreds of neatly arranged binders holding donor data. The director felt this was a nice way to store data and incentivise employees. She explained as follows:

> It wasn't considered an ethical issue. It was an aesthetic issue. Plus, it is difficult to persuade people to donate blood stem cell in China [because

21 Buzan, B., Waever, O. and Wilde, J. 1998, *Security: A New Framework for Analysis*, Boulder: Lynne Rienner Publishers.
22 Zhang, J. Y. 2010, 'China's Regulation of Stem Cell Research in the Context of Cosmopolitanisation', Doctoral Thesis, Department of Sociology, London School of Economics and Political Science.

> traditionally blood is considered an essence of vitality]. I thought a whole wall display of the data books behind the glass panel is a magnificent display of our hard work. It would boost morale... But during my visit to stem-cell donor data banks in the US, I didn't see any data-collection shelves throughout my whole trip. I asked them why. They told me data books are stored in limited-access rooms to protect patients' privacy. Then I realised: Ah! The display of shelves is an ethical issue.

So whilst in the US, arrangement of data books implies protection of patients' confidentiality and professional accountability, in China, it was originally perceived as an 'aesthetic issue'. The director's trip to the US made her think from an alternative perspective: office-space arrangement not only matters to staff members, but also has wider implications to stem-cell donors, patients and medical practitioners. One consequence of the exposure to foreign practice, however, is that it unexpectedly extends the range of ethical-related issues that Chinese stem-cell banks acknowledge.

There is no textbook answer to the complete set of ethical issues that scientists or administrators should be aware of. Instead, what stakeholders have on hand is a growing list of items they need to consider. This list increases as stakeholders' communicative circles expand. The internationalisation of science rewrites the criteria for good practice through the process of ethicisation, or the institutionalisation of a societal issue. Such institutionalisation leads to the encouragement and requirement that scientists and other stakeholders (policymakers, educators, patients) consider a specific concern whilst reflecting on their practice and outcome.

A similar process can be seen in relation to biosafety. Whilst biosafety was considered a personal lab issue prior to the SARS outbreak in 2003 and accidental exposure in 2004, it is now a matter of national — indeed international — security. This has prompted China to pay far more attention to biosafety and pour considerable resources into ensuring ethical conduct within labs. For instance, during a visit to an infectious-disease hospital in Shanghai, Barr was struck by the status and level of respect paid to two young staff members who had recently returned from a World Health Organisation (WHO)-sponsored training programme on biosafety. Although they were quite junior staff, the director of the hospital and another senior colleague indicated that the two employees were in high demand and crucial to the successful running of the site since work in their newly built BSL-4 lab could not proceed without their assistance. Thus, an issue that was previously of local or even personal importance has become institutionalised as a wider ethical and security matter.

One lesson we draw from this is that it is vital to be attuned to how the global or local interface may influence, in any given national context, what type of issues are on the educational agenda (or, crucially, not). In China, where science is a key component of its national development and political and economic strategy, it is imperative that staff are aware of foreign practices and concerns as to what constitutes an ethical or security issue. It is vital for Chinese scientists to be seen to be doing the right thing. This means that a balance must be achieved between advocating an educational agenda, based on international norms, that is 'right' and 'ethical' and, as we shall discuss below, not overselling or pushing foreign solutions onto Chinese partners.

Scientists as Ethics Decision-Makers

After an issue makes its way onto the agenda, a key question then becomes who has the social power and authority to negotiate it? Earlier we discussed the nascent but growing role of bioethicists in China. In line with this, Zhang has found that it is often scientists who are the key players in ethical policy matters but that ethicists have been slowly making progress in getting their voices heard. One respondent put it this way:

> We are entitled to the right of speech. We can express our opinions towards the policy, but that is all. For most major policy, it is still made according to what the officials envisaged. In other words, we can criticise the policy, both through public media and through official administrative channels, but they [government policymakers] don't really listen to all the experts' advice… Of course they listen to the 'hard-core experts' advice, especially for stem-cell research, MOST [Ministry of Science and Technology] won't listen to us [ethicists], it only turns its ear to real scientists.

Another senior bioethicist spoke of her efforts to travel across China in order to raise awareness of the non-scientists' ability to help adjudicate bio problems.

> My role was to 'make a little noise'… Really, what is essential in promoting bioethics in China is, as my colleagues and I call it, 'to make a little noise'… In the past few years, my colleagues and I have been invited by institutions across China to give lectures, give speeches, provide training or write for newspapers. Whenever the topic touches on bio-regulation, we tell people our perspectives on biomedical research. In fact, we are creating general alertness. We try to raise others' attention, make people think about these issues, and see whether we should [have some form of administrative action].

Whilst in many Western countries, bioethics as a discipline, has been professionalised and is part of the institutional framework for debating ethical issues, in China bioethics has shown a degree of separation between its social function and professional merit. On reflecting how Chinese ethicists could improve their social functions, one ethicist at the Chinese Academy of Social Sciences (CASS) said:

> Bioethicists should not be living in a vacuum. I think most of the ethicists [in China] are familiar with Western thinking. In what is considered ethical and what is not, [Chinese] ethicists don't really follow the Chinese convention. As a profession, we are heavily influenced by American and European academia. Especially after our [stem cell] guideline was issued, questions arise from both inside and outside China. We [ethicists] must change our approach towards addressing ethical concerns [in China], and start thinking about these questions. Therefore, we really should examine our own tradition and how it is different from the West... I'm not quite sure how ethical consulting agencies work in the US, how they function, but I do feel that they have a say in policymaking. What is more, they always incorporate a wide range of concerns, religion, science, ethics, law... Their report includes all kinds of opinion from all kinds of social groups. In comparison, when bioethicists are contributing to policymaking in China, I feel we can only propose some personal opinion on this issue. We cannot mobilize a wide range of social groups [to participate]. Public participation is minor and professionals at different levels also have little chance of being involved.

Similarly, during trips to Fudan University in Shanghai and Sun Yat-sen University in Guangzhou, Barr discovered that most, if not all, biosecurity classes and textbooks were being designed and taught without any input from ethicists or social scientists. The only survey that seems to have been conducted into biosecurity educational curriculum in China was organised by a leading microbiologist at Fudan. We discuss some of the possible reasons for the lack of input from social scientists and ethicists below. Furthermore, from our vantage point, the dominant role of scientists as ethics deciders in China implies several key lessons.

Firstly, it was important to know the social and academic status of the actors we wished to reach and what was and was not within their abilities to deliver. In our case, we learned that involving bioethicists was a useful way of getting insights into the larger context of Chinese humanities and social-sciences education but that crucially, if we wanted to make real progress in getting biosecurity modules into the classroom, we needed to collaborate extensively with life scientists. Secondly, when dealing with scientists, we went out of our way to gain their trust by showing that we knew what we were talking about and could relate in some sense to their priorities as leading life scientists. Thirdly, we found it tricky to share a common language with those not well versed in the humanities or social sciences (a problem obviously not unique to China). This meant that we had to be more creative, for example, in discussing why seemingly abstract ethical theories may be appropriate to scientists engaged in dual-use research.

The Status of the Social Sciences

Some of the challenges described above can be traced to the general status of the social sciences (and by association, humanities) in China. We have identified three broad areas in which the history and role of social scientists and ethicists in China may impact efforts to promote bioethical and biosecurity education.

Firstly, there is the question of self-censorship within Chinese academia. CASS was established in 1977 to advise the government and Communist Party and acted as an in-house think-tank for various political factions. It was set up to serve the reforms of Deng Xiaoping and essentially to separate the social from the natural sciences. After the involvement of many CASS staff in the 4 June 1989 demonstrations, the Academy was reformed — or one might say, 'rehabilitated'. Whilst there are ample outlets for CASS academics to publish, many think twice before exposing their work to political criticism. Li Tieying, former President of CASS is rather clear on this point: 'At CASS, behaviour that damages the Chinese Communist Party (CCP), socialist China, the world of social science, or CASS itself is absolutely intolerable'. If, as is sometimes the case, writers do not know where the border lies between safety and punishment, then understandably there is a tendency to 'retreat into a conspiracy of silence', with a reluctance to take on issues which may be seen as sensitive — such as biosecurity.[23]

Secondly, twentieth-century social science in China maintained not only a uniform tendency towards empirical research but also a belief in its technocratic potential to transform society.[24] Despite the effects of the Cultural Revolution (1966–76), when academic inquiry ground to a halt, the social sciences in China today tend to hold similar beliefs. It is only a slight generalisation to say that in China academics often view their work as service to the nation. This is perhaps a difficult concept for Western academics to appreciate. It is impossible in China today to speak against patriotism. The term *youhuan yishi* (忧患意识) is relevant here as it refers to worries about the future of China, common amongst intellectuals.[25]*Youhuan yishi* constitutes a moral concern with improving the nation's wellbeing. It denotes rhetoric of worrying about the nation that remains integral to Chinese intellectual discourse, whether it be interpreted as simply

23 Sleeboom-Faulkner, M. 2007, 'Regulating intellectual life in China: The case of the Chinese Academy of Social Sciences', *The China Quarterly*, vol. 189, p. 99.

24 Chiang, Y. 2001, *Social Engineering and the Social Sciences in China, 1919–1949*, Cambridge: Cambridge University Press.

25 Davies, G. 2007, *Worrying about China: The language of Chinese critical inquiry*, Cambridge: Harvard University Press.

improving the socio-economic plight of the Chinese, or recovering a genuine way of 'being Chinese' and China's international status as a great scientific and cultural power.[26] One consequence of this is that amongst Chinese intellectuals (that is, some of the key academics, whose support is necessary to help further biosecurity awareness), there is ambivalence towards foreign or imported ideas.

Thirdly, one consequence of the state of social sciences in China is that there are sometimes no comparable fields to the ones we tend to take for granted in the West. For example, there is no real equivalent in China to Science and Technology Studies. Thus, some of the academic terms, theories, concerns, and norms that we use do not necessarily carry over. In the West, for instance, many working along the science/society/security interface subscribe to the notion of the co-production of knowledge — that is, to the belief that the natural and social order are produced together and the ways in which we know and represent the world are inseparable from the ways we chose to live in it. According to this view, science does not mirror reality. Rather, it both embeds and is embedded in social practices, norms, identities, and institutions.[27] Such a school of thought may seem out of place in China. Gloria Davies who writes that 'Unlike its Western counterpart, Chinese academic discourse remains a quest for certainty, a quest envisaged in terms of the acquisition of better, more rational, and theoretically refined knowledge', captures this difference nicely.[28]

All of this may seem far removed from bioethics and biosecurity education. But there are several lessons we take from these points. In one recent case, Barr sought funding to undertake a project on biosecurity in China. The funder stipulated that Chinese partners must come from CASS. However, it emerged that there was no one at the Academy who had relevant expertise for the application.[29] Although we cannot know for certain, it seems possible that one reason this area is avoided is its political sensitivity; that is, its proximately to questions of national security. During an earlier experience in China, Barr was told by a senior ethicist that conducting ethics research into infectious disease would be very sensitive and that it would be necessary to proceed cautiously since some people may not be willing to talk about the topic given its connection to matters of security and government censorship.

It is interesting here to consider the official oath taken by medical students. In addition to the usual duties of moral discipline and medical skill, it includes mention of love of country and people:

---

26 Link, P. 1998, *Evening chats in Beijing*, New York: W.W. Norton and Company, Inc.

27 Jasanoff, S. 2004, *States of knowledge: The co-production of science and the social order*, London: Routledge.

28 Davies 2007, op. cit.

29 In a way this may not be surprising since, as discussed above, biosecurity is new to the policy agenda. Yet despite being 'new', it has not stopped some life scientists from getting involved with biosecurity work.

> I will devote my life to the cause of medicine. I love my country. I will be loyal to the people. I will adhere to strict medical ethics. I will respect my teachers and observe discipline. I will study diligently and constantly in order to perfect my medical skills and to develop myself in all possible aspects — morally, intellectually and physically.[30]

The point here is that, in China at least, there is sometimes a strong (techno) nationalist backdrop to many issues, especially ones such as the science/society/security interface. This is important, in part, since the government is by far the largest funder for scientists. Seen this way, promoting a patriotic discourse may serve several purposes. One key strategy the government has used to lure scientists back to China is to appeal to their sense of nationalism in helping to build a scientifically competitive superpower. This relates to a final point worth considering for foreign researchers wishing to collaborate with Chinese partners. It has to do with earning trust — or, to put it another way, avoiding suspicion. As we went about our separate studies, it was not uncommon for us to be asked whether or not we were journalists. This fear seemed to stem not so much from concerns over security but rather that we would do or print something to harm our respondent's (and China's) reputation. Some the most damaging stories about hospitals, doctors, and ethics violations have sprung from the media in China. As we conducted our work in the run up to the 2008 Olympics, we could not help but notice a worry over being embarrassed whilst the international spotlight shone on the Beijing Games. Of course, it is always the case that researchers must work in partnership with each other. Our point here is for foreign researchers working in China to be sensitive to the patriotic backdrop to Chinese science, an element that is often missing in Western academia.

Conclusion: What Can Be Done?

There are numerous ways to help promote biosecurity education in China. We conclude this chapter by suggesting three key interrelated contributions that researchers (Chinese and foreign) may make.

First, there is a need to improve data sharing and cross-lab communication in China. Our work has found that whilst scientists in China tend to talk about collaboration with other local sites, in fact, communication between labs is more limited than one might expect.[31] The first step here is to help promote the type of workshop that CAS ran in 2008 on dual-use research (cited above).

30 People's Republic of China Ministry of Education 1991, *Official Oath for Medical Students*, Appendix 4 of Document No. 106.

31 Zhang, J. Y. 2010, 'The organization of scientists and its relation to scientific productivity: Perceptions of Chinese stem cell scientists', *BioSocieties*, vol. 5(2).

The goal of such a meeting is not only for biosecurity-minded staff in China to exchange views with foreign experts. Whilst this is important, there is also a great need for researchers within China to share and learn from one another. China is in the unique position of having the second-largest economy in the world and yet in per-capita income terms, it ranks ninety-ninth.[32] This shows just how uneven growth and development has been, which has knock-on effects for what different sites will consider to be the range of biosecurity issues and priorities it faces. Of course income gaps will also have significant effects on levels of resources and abilities to promote biosecurity awareness. Thus, it is crucial that scientists from China's leading centres are brought together with staff from less-resourced hospitals and labs. Only in this way can we ascertain across sites what is already being done, at what levels, and for which students and lab workers. This leads us to our second point, which is to use this data to help build biosecurity capacity in China.

Second, biosecurity-minded researchers ought to work towards establishing compulsory biosecurity classes for all life-science students. This could supplement scientific-morality courses that are already required of most students. It is sometimes mistakenly assumed that in China all change comes from the top down. Whilst government statements at the BTWC illustrate the political importance of biosecurity, our experience is that the fastest way to change is when leading institutions — such as CAS — make grass-roots initiatives that, once established, draw the attention and support of government ministries. The aim here is to work with interested staff at high-profile key sites to assess the possibilities and practical issues that must be overcome in establishing a required course (including instructors who are trained to deliver such a class). These moves can then be supported with biosecurity and dual-use teaching resources appropriate to the needs and context of the life sciences in China. This leads to our third point.

Third, in order to help facilitate mandatory courses, appropriate materials are needed. One way to do this would be to draw on the EMR, an open-access series of lectures to support life scientists and educators in learning about biosecurity and dual-use issues (see the chapter by Whitby and Dando). The aim here is to help interested social and life scientists in China design resources and methods suitable to their context (with, for example, particular emphasis on the security implications of infectious-disease research). This suggestion requires researchers working together, across sites, to develop and share texts and ideas for the effective delivery of biosecurity classes. It would not necessarily require starting from scratch, as part of the aims in our first suggestion would be to assess which

32 International Monetary Fund 2009, *World Outlook Economic Database*, Washington: International Monetary Fund.

sites are using which materials and provide the chance for biosecurity educators to reflect with one another on the usefulness or otherwise of the books and examples they employ.

If adopted, these suggestions could have positive knock-on effects. One of the main regulatory challenges in China is not a lack of rules but rather, a lack of effective enforcement of the rules that already exist. Thus, our suggestions are practical steps that aim to help create a culture of responsibility, whereby enforcement mechanisms are strengthened through widespread awareness and education.

Chapter 7: Raising Awareness among Australian Life Scientists

CHRISTIAN ENEMARK

Introduction

In early 2001, the year of the anthrax envelope attacks in the US, research conducted by a group of Australian scientists highlighted the security implications of the dual-use dilemma in the life sciences. This group was attempting to produce an infectious contraceptive for mice, which periodically breed out of control in parts of Australia. The scientists first spliced the *zona pellucida glycoprotein 3* (ZP3) gene into a mild mousepox virus in the hope of inducing antibodies with a contraceptive effect.[1] They subsequently inserted the *interleukin-4* (IL-4) gene, which helps regulate immune system reactions, to boost this genetically engineered sterility treatment. However, the IL-4 gene increased the virulence of the virus such that it rapidly killed normally resistant mice. The researchers subsequently showed that the expression of IL-4 resulted in a strain of mousepox so powerful that it killed even vaccinated mice.[2] A disturbing implication of this finding is that adding an IL-4 gene might similarly increase the fatality rate of smallpox (or some other poxvirus that infects humans) and potentially allow the virus to circumvent vaccination. The Australian group's findings were published in the peer-reviewed *Journal of Virology*, but they also attracted attention in the popular magazine *New Scientist* as well as in the mainstream media.[3] Ethical questions posed to this experience included: Should this research have been done? Should the results have been published? How

1 Jackson, R. J., Maguire, D. J., Hinds, L. A. and Ramshaw, I. A. 1998, 'Infertility in mice induced by a recombinant ectromelia virus expressing mouse zona pellucida glycoprotein 3', *Biology of Reproduction*, vol. 58(1), pp. 152–9.

2 Jackson, R. J., Ramsay, A. J., Christensen, C. D., Beaton, S., Hall, D. F. and Ramshaw, I. A. 2001, 'Expression of mouse interleukin-4 by a recombinant ectromelia virus suppresses cytolytic lymphocyte responses and overcomes genetic resistance to mousepox', *Journal of Virology*, vol. 75(3), pp. 1205–10.

3 Nowak, R. 2001, 'Killer virus', *New Scientist*, 10 January, available: http://www.newscientist.com/article/dn311 [viewed 23 February 2010]; Broad, W. J. 2001, 'Australians Create a Deadly Mouse Virus', *New York Times*, 23 January, available: http://www.nytimes.com/2001/01/23/health/23MOUS.html?pagewanted=all [viewed 23 February 2010].

should scientists and governments respond to dual-use dilemmas? This chapter explores recent efforts by four Australian academics, one of whom was involved in the mousepox IL-4 research, to facilitate discussion among life scientists of these and related questions.

In 2009, the US-based Alfred P. Sloan Foundation funded a pilot series of four interactive seminars for Australian scientists and students on the potential security risks of laboratory research on pathogenic micro-organisms. The seminars were designed and facilitated by a team of academics from the National Centre for Biosecurity (NCB), a collaboration of the University of Sydney and the Australian National University (ANU). This project was part of a multinational programme of education and awareness-raising on the dual-use dilemma in the life sciences, coordinated by Brian Rappert of the University of Exeter (UK). The Australian team was multidisciplinary and consisted of virologists Belinda Herring (Sydney) and Ian Ramshaw (ANU), bioethicist Michael Selgelid (ANU), and political scientist Christian Enemark (Sydney). We had sought funding to conduct these seminars because we perceived that the issue of the dual-use dilemma in the life sciences had received scarce attention in Australia at the level of higher education and professional training. In 2005 the Australian government had acknowledged, at a meeting of Biological and Toxin Weapons Convention (BTWC) member-states in Geneva, that 'Amongst the Australian scientific community, there is a low level of awareness of the risk of misuse of the biological sciences to assist in the development of biological or chemical weapons'.[4] Over the course of 2009, we gained the impression that little had changed since that observation was made.

At the level of government policy, Australia had recently introduced a scheme under the *National Health Security Act 2007* to regulate the handling, storage, transfer and disposal of 'Security-Sensitive Biological Agents' (SSBAs). A government-run 'road show' of workshops in major Australian cities in 2008–09 sought to inform 'affected stakeholders' about the SSBAs scheme. As part of this, the 'dual-use dilemma' was mentioned briefly at the start of each workshop.[5] After attending one of these, our impression was of a top-down, rulers-to-the-ruled, regulations-oriented approach to awareness-raising. We felt the time was ripe for some bottom-up, non-official, ethics-oriented engagement with Australian life scientists and students on the challenge of preventing the

4 Meeting of the States Parties to the Convention on the Prohibition of the Development, Production and Stockpiling of Bacteriological (Biological) and Toxin Weapons and on their Destruction, Meeting of Experts, *Raising awareness: Approaches and opportunities for outreach (working paper prepared by Australia)*, BWC/MSP/2005/MX/WP.29, Geneva, 21 June 2005: 1.

5 Department of Health and Ageing 2010, *2009 SSBA regulatory scheme road show*, available: <http://www.health.gov.au/ssba#roadshow> [viewed 20 January 2010].

destructive use of science. Our hypothesis was that an interactive seminar on the dual-use dilemma, facilitated by academics, would make for a more relaxed and fruitful forum for raising questions and discussing concerns.

Preparation

Our team of facilitators prepared for the 2009 seminar series by undertaking two days of training in December 2008. Rappert and his US colleague Nancy Connell (University of Medicine and Dentistry of New Jersey) visited the University of Sydney and we spent the first day airing our views on the dual-use dilemma and exploring issues of seminar content and awareness-raising methodology. The following day, the team and its trainers jointly conducted a two-hour demonstration seminar on campus involving around 20 participants from a variety of academic and government backgrounds. This was an opportunity for us not only to get a feel for an interactive seminar on the dual-use dilemma but also obtain feedback from participants and each other on matters of substance and style. Four weeks prior, we had distributed an invitation flyer via email to relevant university mailing lists. The flyer was reused for all four seminars in 2009 and was worded as follows:

> International concern about biological weapons has increased significantly, particularly since the anthrax attacks of 2001 in the United States. Biotechnology research has great potential to benefit health and agriculture, but questions arise as to whether it might aid the deliberate spread of disease. Traditional concerns about laboratory biosafety are being increasingly complemented by growing attention to biosecurity practices within and outside of laboratories. In Australia, this is reflected in the 'security-sensitive biological agents' (SSBAs) scheme to be introduced in 2009. Beyond legal regulations, ethical questions remain including what novel threats might stem from life science research, whether and how scientists should contribute to national defence, and whether some lines of investigation are too contentious to pursue.

This interactive seminar has three aims:

1. Inform participants about current international discussions surrounding 'dual use' and 'biosecurity'.
2. Generate debate about the merits and pitfalls of proposed policy responses.
3. Provide examples of educational programmes and oversight measures related to dual-use research.

> Topics for discussion will include the funding of research, communication of research results, oversight of experiments, the responsibilities of scientists and other biosecurity stakeholders, and examples of various national and international measures being implemented or considered.

The flyer appeared on NCB letterhead featuring the University of Sydney and ANU logos and the NCB website URL (www.biosecurity.edu.au), and it acknowledged funding from the Alfred P. Sloan Foundation. We wanted to make clear to prospective participants that the seminar was academic in nature and not associated with government in any way. Our intention in so doing was to distinguish the seminar from the SSBAs 'road show' passing through Australian cities in 2008–09.

Participants at the December 2008 demonstration seminar were reminded at the outset that it was being conducted primarily for training purposes. After a brief introduction to the problem of biological weapons and the awareness-raising rationale for the following year's seminar series, Rappert spent one hour taking participants through a seminar similar to those that he and his British colleague Malcolm Dando (University of Bradford) had been conducting with scientists worldwide since 2004.[6] Rappert's content, structure and style were the model for the seminars we conducted in 2009. Connell and the Australian team of facilitators then spent around 50 minutes asking participants to expand on points they had raised earlier. We concluded the demonstration seminar by eliciting feedback, thanking our guest facilitators and the participants and then inviting everyone to join us for lunch. Immediately afterwards, Rappert and Connell debriefed us on possible reasons for the success or otherwise of different aspects of the seminar. Taking these lessons on board, and building on our collective research and teaching experience as academics, we were now trained and ready to conduct discussions around Australia on the dual-use dilemma in the life sciences.

The Seminars

Over the course of 2009 we undertook seminars at the John Curtin School of Medical Research at ANU (20 March), the Department of Microbiology and Immunology at the University of Melbourne (9 June), the Centre for Infectious Diseases and Microbiology at the Westmead Hospital campus of the University of Sydney (23 July), and the Queensland Institute of Medical Research in Brisbane (17 August). Each of these took place over a one-hour period and on average attracted around 50 scientists and students. Many of the participants

6 Rappert, B. 2009, Experimental secrets: International security, codes, and the future of research, New York: University Press of America.

in the December 2008 demonstration seminar had provided feedback that even two hours was not enough time. However, in the interests of maximising participation, we decided it would be better to have our seminar included in existing programmes of weekly one-hour seminars which scientists and students are in the habit of attending. Given limited time and a desire to avoid 'death-by-PowerPoint', we settled on 11 slides:

1. Title slide ('The Dual-Use Dilemma in the Life Sciences') including NCB banner, names and university affiliations of seminar facilitators and name of sponsor.
2. Three questions:

   What research should or should not be done?

   How should research results be communicated?

   What forms of security oversight, if any, are required for life scientists?
3. What research should or should not be done?

   Example: synthetic poliovirus
4. How should research results be communicated?

   Example: resurrected 1918 flu virus
5. What forms of security oversight, if any, are required for life scientists?

   Example: laws and codes of conduct
6. 1972 Biological and Toxin Weapons Convention

   Text of Article 1
7. National Health Security Act 2007

   List of security-sensitive biological agents
8. A Code of Conduct for Biosecurity (Royal Netherlands Academy for Arts and Sciences (RNAAS) 2007)

   Basic principles and target groups
9. A Code of Conduct for Biosecurity (RNAAS 2007)

   Research and publication policy
10. A Code of Conduct for Biosecurity (RNAAS 2007)

    Accountability and oversight
11. Thanks and contact information

The first task was for the team of seminar facilitators to introduce themselves and their academic backgrounds. Some participants in the December 2008 demonstration seminar had indicated in feedback that they would also have liked everyone else in the room to introduce themselves. The reason given was typically that this would reveal 'who people are and where they are coming from'. Nevertheless, we decided against this, partly because it would take up too much time and most seminar attendees at a given research institution would probably already know each other, but mainly because we felt that people would speak more freely and frankly (to us) under conditions of relative anonymity.

After the facilitators' introductions, we briefly explained why the seminar was being conducted. The 'dual-use dilemma' is nothing new, we assured the participants, and can arise in any and every branch of science. Simply stated, it is the notion that the fruits of scientific endeavour can be used for purposes both good and bad. But whereas the downsides of nuclear science, for example, have been comprehensively canvassed, there has hitherto been little open discussion of the dual-use dilemma and biological weapons risks among life-science professionals. This is despite the fact that the Australian Academy of Science, along with 67 other academies worldwide, has endorsed the 2005 *Statement on Biosecurity* by the Interacademy Panel on International Issues. The statement acknowledges that 'some science and technology can be used for destructive purposes as well as for constructive purposes', and among the principles it offers to guide individual scientists is that 'scientists should be aware of, disseminate information about and teach national and international laws and regulations, as well as policies and principles aimed at preventing the misuse of biological research'.[7] We explained that this was a challenge around the world, and that our Australian series of seminars in Canberra, Melbourne, Sydney and Brisbane was roughly mirroring the work of seminar teams based in the UK, South Africa, Israel, Japan and the US. Our seminar, unlike those the participants were used to attending as part of their institution's weekly programme, was not about presenting research findings and nor was it a research exercise in itself.

After setting the scene, we briefly explained how we wanted the seminar to run. While pointing to the political reality of government concern about the problem of biological weapons, it was not our role to make judgments about if or the extent to which this concern is justified. Nor was it the purpose of the seminar, we explained, to advocate specific practices and policies. Rather, we wanted to provide Australian life scientists and students with a forum for structured discussion among themselves of the dual-use dilemmas associated with their work. In any event, we the seminar facilitators were ourselves divided

---

7 Interacademy Panel on International Issues 2005, *Statement on biosecurity*, available: <http://www.interacademies.net/Object.File/Master/5/399/Biosecurity%20St..pdf> [viewed 9 January 2010].

on possible solutions to dual-use dilemmas. Our objectives of education and awareness-raising would be best achieved if the seminar participants did most of the talking over the hour. We then indicated our intention to allow roughly equal time for participants to discuss each of three questions:

- What research should or should not be done?
- How should research results be communicated?
- What forms of security oversight, if any, are required for life scientists?

These questions are along the same lines as those that Rappert and Dando have identified as central to current dual-use knowledge debates.[8] A different facilitator would initiate discussion of each question with a real example of a dual-use dilemma, then it would be up to the seminar participants to talk through the issues as they saw them and in the light of any comments we made. At the first two seminars (Canberra and Melbourne), it was difficult to get someone in the room to be the first to talk about each question. On the advice of a Melbourne seminar participant who had lunch with us afterwards, we later employed the device of first asking for a show of hands. This quickly served to demonstrate to each participant that others in the room also had an opinion and, more importantly, that opinions differed.

To stimulate discussion on our first question (what research should or should not be done?), we referred to an experiment that involved synthesising the genome of a pathogen. The dilemma is:

1. Synthesis of the genomes of viruses theoretically allows the introduction of mutations or novel sequences that can be used to study the function of particular genes or regulatory sequences.
2. Synthesis technology would obviate the need to source pathogens from natural reservoirs in other parts of the world or from other laboratories. It could also facilitate recotnstruction of extinct pathogens and enable construction of novel pathogens.

Our example was a project in which US scientists sponsored by the US Department of Defense spent three years synthesising the 7500 chemical units of *poliomyelitis* (polio) virus. Referring to the published polio virus RNA genome, they strung together corresponding DNA sequences purchased over the Internet. This was used in a cell-free extract to create live virus that paralysed and killed mice. The results, published in 2002, showed that eradicating a virus in the wild might not mean it is gone forever.[9] We asked the seminar participants: Should

---

8 Rappert, B. 2007, 'Education for the life sciences' in Rappert, B. and McLeish, C. (eds), *A web of prevention: Biological weapons, life sciences and the future governance of research*, London: Earthscan, p. 60.

9 Cello, J., Paul, A. V. and Wimmer, E. 2002, 'Chemical synthesis of poliovirus cDNA: generation of infectious virus in the absence of natural template', *Science*, vol. 297(5583), pp. 1016–8.

this research have been done? In general, responses to this question tended towards preferring a permissive approach, but there was some variation from one institution to the next. In Brisbane, an initial show of hands indicated a majority were in favour of the experiment having been done, but most Sydney participants adopted a neutral position. Nobody offered the view that polio virus synthesis was particularly groundbreaking or important work. Rather, as one Brisbane participant observed, 'This was not research, it was just applying known technology'. A Melbourne participant posed the rhetorical question: 'why would you bother?' and in Sydney one person likewise suggested that the experiment was 'pointless' but otherwise 'morally neutral'.

At this point the seminar discussion would usually quickly turn, as we anticipated, to a more general consideration of scientists' freedom to experiment. A common theme was that no experiment exists (or can be judged) in isolation, but rather builds on research that has gone before; there is a general backdrop of pre-existing scientific knowledge and technological capabilities. To test this view, we would offer a proposition along the lines of 'But what if one research project was what turned a reasonably foreseeable, dangerous but theoretical possibility into a real danger?' A Canberra participant commented that 'evil applications of science cannot always be predicted' and an attitude prevalent throughout the seminar series was that shutting down lines of research might deprive humanity of important, life-saving discoveries. On this point, one of the more interesting moments during the Melbourne seminar was when one participant, an influenza researcher, pointed to work currently being carried out which involves increasing the transmissibility of a pathogen. The dilemma is:

1. For public-health planning purposes, it may be important to know whether a naturally occurring infectious-disease threat could be worsened by the evolution of a pathogen into a more transmissible form.
2. A pathogen might be more useful for biological weapons purposes if it is more easily transmitted through a target population.

The World Health Organization is presently sponsoring research to find out whether the H5N1 avian influenza virus could mutate to produce a human influenza pandemic. There is hope that, by re-assorting (mixing) H5N1 with human influenza viruses in the laboratory, scientists may determine how dangerous the hybrid virus would be and the likelihood of it causing a pandemic. One such experiment in re-assortment has found that transmissibility is not increased.[10] However, a successful experiment of this kind could result in a man-made pandemic influenza virus that would need to be kept extremely secure against theft or leakage from the laboratory.

From time to time, some participants appeared reluctant to mesh scientific and ethical concerns. One pragmatic pronouncement made at the Brisbane seminar was that 'if I don't do it, someone else will, unless you inhibit us all'. This then prompted the observation that 'any restriction not globally applied has no value'. At all the seminars, there invariably arose a question along the lines of 'Who decides at which point something is too dangerous?' In reply, one of the facilitators would ask participants their views on a hypothetical experiment involving chemical synthesis of *variola* (smallpox) virus, a pathogen which no longer exists in nature. There was some disagreement over the technical feasibility of producing a virus capable of infecting and sickening humans, but no one argued that there would be much scientific merit in doing so. On the question of whether such work should or should not be done, as a matter of morality, one Melbourne participant replied by asking the group: 'Hang on, is this a scientific or a social issue?' This in turn received the reply 'Others will say it is a social issue, even if we don't think so.'

Attention then turned to what might be considered the lifeblood of science — information and its sharing. The second question we posed was closely related to the first: a person's opinion on whether or how to communicate research results is likely to be impacted by whether they think the research should have been done at all. To stimulate discussion, we referred to a pair of experiments that centred on genetic sequencing of a pathogenic micro-organism. The dilemma is:

1. Sequencing the genetic code of entire pathogens or specific genes of pathogens could assist in understanding the nature of the pathogen and in the development of new vaccines or treatments for the disease it causes.
2. Gene-sequence data could be used to reconstruct a pathogen (or one with its harmful characteristics) for deployment against a target population with no natural immunity.

Our example was research results, published in 2005, on the complete genetic sequencing of the 1918 influenza A (H1N1) virus and the resurrection thereof using reverse genetic techniques.[11] This revealed (and reproduced, in animals

10 Maines, T. R., Li-Mei Chen, Matsuoka, Y., Chen, H., Rowe, T., Ortin, J., Falcón, A., Nguyen, T. H., Le Quynh Mai, Sedyaningsih, E. R., Syahrial Harun, Tumpey, T. M., Donis, R. O., Cox, N. J., Subbarao, K. and Katz, J. M. 2006, 'Lack of transmission of H5N1 avian–human reassortant influenza viruses in a ferret model', *Proceedings of the National Academy of Sciences*, vol. 103(32), pp. 12,121–6.

11 Taubenberger, J. K., Reid, A. H., Lourens, R. M., Wang, R., Jin, G. and Fanning, T. G. 2005, 'Characterization of the 1918 influenza virus polymerase genes', *Nature*, vol. 437(7060), pp. 889–93; Tumpey, T. M., Basler, C. F., Aguilar, P. V., Zeng, H., Solorzano, A., Swayne, D. E., Cox, N. J., Katz, J. M., Taubenberger, J. K., Palese, P. and Garcia-Sastre, A. 2005, 'Characterization of the reconstructed 1918 Spanish influenza pandemic virus', *Science*, vol. 310(5745), pp. 77-80.

at least) the traits that made the pandemic influenza virus so virulent (the 'Spanish flu' killed around 50 million people). However, the publication of this information gave rise to concerns that would-be bioterrorists could use it to reconstruct this strain of H1N1 for malign purposes. The US National Science Advisory Board for Biosecurity, a non-government advisory body, was asked to consider the relevant papers before publication (in *Science* and *Nature*) and concluded that the scientific benefit of the future use of this information on the 1918 virus far outweighed the potential risk of misuse.[12] We asked the seminar participants: 'Was publication the right thing to do?'

The overwhelming majority favoured publication, and the ensuing discussion typically included appeals to scientific values. As scientific discovery is a cumulative process, as one Canberra participant observed, it is 'essential to provide [publish] methodology so it [the experiment] can be replicated'. Another argument frequently advanced was one that challenged the premise of our second question: 'You can't control publication; we've got the Internet.' In Canberra this elicited the reply 'but you can control publication in a commercial or military context', and one Sydney participant said 'I'm more worried about results not being published because of government classification.' The implication was that governments themselves could secretly be using life-science technology for malign purposes, or that classification could be hindering efforts to protect the community at large. On the latter point, a Brisbane participant argued that 'publication raises awareness and makes us more prepared'. As a way of testing the assumptions underlying the fiercely pro-publication sentiments expressed at all the seminars, we would ask: 'Can you imagine any research results that should not be communicated at all?' At the Melbourne seminar, one participant replied with the hypothetical example of 'accidentally' discovering a method of making HIV transmissible as an aerosol. Under no circumstances, he argued, would he seek to place that method in the public domain.

Our third question (What forms of security oversight, if any, are required for life scientists?) invited participants to critically assess existing and proposed governance mechanisms, whether grounded in specific laws (for example, biosecurity regulations) or ethical principles (for example, professional codes of conduct). Whereas the previous two questions had canvassed possible problems associated with pursuing particular lines of research and communicating results, participants were now offered an opportunity to exchange ideas in a solutions-oriented frame of mind. In this third and final part of the seminar, our aim was to shift life scientists' attention away from the relatively familiar terrain of laboratory techniques and research publication and towards the more exotic realm of national security concerns. As an exercise in awareness-raising,

---

12 Miller, S. and Selgelid, M.J. (2008) *Ethical and philosophical consideration of the dual-use dilemma in the biological sciences*, Springer, p. 26.

we set about acquainting participants with the existence and content of some relevant biosecurity laws as well as a sample code of conduct. It was not our role, however, to champion legal or ethical oversight *per se* or any specific mode thereof. It was up to those attending the seminar to discuss among themselves whether particular oversight mechanisms were feasible and/or desirable and why.

Turning first to the issue of legal regulation, we framed the security dilemma in the life sciences as follows:

1. Biological weapons threats can emanate from trusted laboratory personnel, even those with high-level security clearances.
2. Governments need to manage the risk of imposing too great a regulatory burden. A reduction in potentially life-saving research, precipitated by scientists opting out of laboratory work, could undermine capacity to resist both natural infectious-disease outbreaks and biological attacks.

We illustrated this dilemma by briefly referring to two recent episodes from the US with which a minority of participants appeared to be familiar. The first example concerned a *Yersinia pestis* (plague) bacteria expert, Thomas Butler. Formerly chief of the infectious-diseases division at Texas Tech University, Butler was the first US scientist to be put on trial for biosecurity offences under post-9/11 legislation. In the US, federal laws regulate the storage, handling, transfer and disposal of 82 named Biological Select Agents and Toxins (BSATs). Butler faced charges including that he smuggled (that is, transferred without permission) *Y pestis*, an agent on the BSATs list. In December 2003 a Texas jury found Butler guilty on three charges related to the shipment of samples to a Tanzanian researcher without the proper permit, the package having been labelled merely as 'laboratory materials'.[13] On 10 March 2004 he was sentenced to two years in prison. The judge had cut seven years off a possible standard nine-year sentence set by federal guidelines, citing testimony that the bacteria shipment was done for humanitarian reasons and that the US Department of Commerce would have approved a transportation permit had Butler applied for one. The judge also cited Butler's early work on treatment of diarrhoeal diseases and oral rehydration as having 'led to the salvage of millions of lives throughout the world'.[14]

The second example concerned a *Bacillus anthracis* (anthrax) bacteria expert, Bruce Ivins. A microbiologist employed for 28 years at a US Government laboratory, Ivins committed suicide on 29 July 2008 before he could be

---

13 Chang, K. 2004, 'Scientist in plague case is sentenced to two years', *New York Times*, 11 March, p. A18; Piller, C. 2003, 'Plague expert cleared of serious charges in bioterror case', *Los Angeles Times*, 2 December, p. A16.

14 *United States v Butler*, 5:03-CR-037-C, US District Court, Northern District of Texas, 10 March 2004, available: <http://www.fas.org/butler/sentence.html> [viewed 7 January 2010].

charged in connection with the anthrax envelope attacks of 2001.[15] Ivins was a published expert on anthrax vaccines,[16] and in 2003 had received the US Defense Department's highest civilian honour for his work in this area.[17] Affidavits for search warrants published by the Federal Bureau of Investigation (FBI) in August 2008 included a description of Ivins's job at the US Army Medical Research Institute for Infectious Diseases in Maryland. One of his tasks was to prepare 'large batches' of aerosolised anthrax bacteria. Animals were then subjected to 'aerosol challenges' to test the effectiveness of vaccines. Ivins knew how to use devices such as lyophilisers, incubators and centrifuges which are 'considered essential for the production of the highly purified, powdered anthrax used in the Fall 2001 mailings'.[18] The picture that emerged from published case documents was that Ivins knew how to make and use biological weapons, and the FBI's scientific and documentary evidence indicating that he did so in 2001 was compelling.[19]

By this stage of the seminar it was easy to elicit opinions, some strongly worded, on what were commonly regarded as shocking real-life experiences of lost innocence in the life sciences. One participant at our Brisbane seminar knew many of the details of Thomas Butler's case and expressed outrage at what he perceived to be an injustice. In Sydney, by contrast, several participants expressed views along the lines of 'he was unlucky; these days you just have to be more careful'. There appeared to be less awareness of, or willingness to talk about, Bruce Ivins, although some participants knew enough to be able to voice doubts about the scientific foundations of the case against him. When this issue was raised at two of the seminars, the facilitators did not express a view but instead pointed to the FBI's decision in May 2009 to commission an independent review of its investigation by the US National Academy of Sciences.[20]

After allowing time for some discussion of the above two criminal cases, the facilitators briefly explained the relevant international legal context and then drew the focus of attention back to Australia. At all four seminars, almost all the participants seemed unaware of the 1972 Convention on the Prohibition of the Development, Production and Stockpiling of Bacteriological (Biological) and Toxin Weapons and on Their Destruction (the Biological and Toxin Weapons Convention or BTWC). However, out of concern that delving too deeply

15 Johnson, C. 2008, 'US settles with scientist named in anthrax cases', *Washington Post*, 28 June, p. A01.

16 Hewetson, J. F., Little, S. F., Ivins, B. E., Johnson, W. M., Pittman, P. R., Brown, J. E., Norris, S. L. and Nielsen, C. J. 2008, 'An *in vivo* passive protection assay for the evaluation of immunity in AVA-vaccinated individuals', *Vaccine*, vol. 26(33), pp. 4262–6.

17 Dance, A. 2008, 'Death renews biosecurity debate', *Nature*, vol. 454(7205), p. 672.

18 Dellafera, T. F. 2008, *Affidavit in support of search warrant*, US Department of Justice, available: <http://www.usdoj.gov/amerithrax/07-524-M-01%20attachment.pdf> [viewed 13 August 2008].

19 Johnson, C., Leonnig, C. D. and Wilber, D. Q. 2008, 'Scientist set to discuss plea bargain in deadly attacks commits suicide', *Washington Post*, 2 August, p. A01.

20 Shane, S. 2009, 'F.B.I. to pay for anthrax inquiry review', *New York Times*, 7 May, p. A25.

into legal provisions might be a turn-off for a scientific audience, we felt it was enough simply to display the text of Article 1. The Convention's in-built recognition of the dual-use dilemma is highlighted in italics:

> Each State Party to this Convention undertakes never in any circumstances to develop, produce, stockpile or otherwise acquire or retain:
>
> (1) Microbial or other biological agents, or toxins whatever their origin or method of production, of types and in quantities that have *no justification for prophylactic, protective or other peaceful purposes;*
>
> (2) Weapons, equipment or means of delivery designed to use such agents or toxins for hostile purposes or in armed conflict.

We explained that the BTWC applies to the actions of states rather than individuals, but that the wording of the above provision is reproduced in Australian criminal law. The maximum penalty for an individual convicted of developing, producing, stockpiling, acquiring or retaining biological agents and toxins for a non-peaceful purpose is life imprisonment.[21] Again, most seminar participants appeared to be unaware of this. However, they were generally much more familiar with the list of 22 SSBAs which we displayed next. Supplementing the general prohibition on bad intentions contained in international and domestic law, the *National Health Security Act 2007* authorised the Australian government's recent introduction of detailed rules for the day-to-day activities of those with access to and knowledge of hazardous pathogens. The SSBAs scheme, although similar to the US BSATs scheme (of which Thomas Butler fell foul) that dates from the mid-1990s, is virtually unprecedented in Australia. Of particular interest to seminar participants was the requirement that persons authorised to handle the most dangerous (Tier 1) SSBAs must undergo a National Criminal History check by the Australian Federal Police and a Politically Motivated Violence check by the Australian Security Intelligence Organisation.[22] It was not our role to explain, attack or defend the SSBAs scheme. Rather, we wanted each of our seminars to be an opportunity for Australian life scientists and students to engage in frank discussion of a clear and present mechanism of security-oriented oversight. Opinions varied as to the wisdom and effectiveness of biosecurity regulations in general. At the Sydney seminar, for example, one participant warned that 'gene-technology regulations [dating from 2000] did impede research' but another said that 'sometimes you have to legislate to bring about changes in behaviour'. A laboratory manager in Sydney observed that 'younger scientists are more likely to adhere to biosafety rules' but additional biosecurity rules would bring, according to a Brisbane

21 *Crimes (Biological Weapons) Act 1976*(Cth), s 8.

22 Department of Health and Ageing 2009, *SSBA Standards, 3.3 Authorised persons*, available: <http://www.health.gov.au/ssba#standards> [viewed 11 January 2010].

participant, 'compliance overload'. Another opinion voiced in Brisbane was: 'It's like Aboriginal housing; layers of bureaucracy make it impossible to get things done.' Towards the end of the Sydney seminar, one participant set heads nodding around the room when she warned, 'Don't take away from scientists that love of what they do.'

As many participants were aware, the Department of Health and Ageing runs a nationwide series of workshops for the purpose of instructing life scientists and laboratory managers on the requirements for complying with the SSBAs scheme. However, these workshops are not an opportunity to suggest changes to the rules or challenge the scheme as a whole. As disinterested academics, we were determined through our seminars to facilitate such an opportunity, and participants readily gave voice to a wide array of opinions. These ranged from outright opposition to the SSBAs scheme on the grounds that it inhibited research through to concerns that the scheme was entirely necessary but not strict enough. Declining to express personal views for or against specific modes of security oversight, we suggested that interactive seminars such as ours could be a valuable, empowering and bottom-up means of increasing the security consciousness of life scientists and students. Arguably, this complements rather than conflicts with the top-down regulatory scheme imposed by the *National Health Security Act 2007*.

Further to the issue of security-oriented oversight, the law does not (and perhaps could not) offer guidance for life scientists making decisions about lines of research (Question 1) and communicating research results (Question 2). These are best regarded as ethical rather than legal issues and, as one Canberra seminar participant commented, 'often it is better to think about moral obligations than regulations'. One mechanism of ethics-based oversight, professionally binding but falling short of actual law, is a code of conduct. The final part of the seminar, drawing all its themes together with possible solutions in mind, invited participants to consider sample principles for a code. We used the example of the 2008 document *A Code of Conduct on Biosecurity* commissioned by the Royal Netherlands Academy of Arts and Sciences at the request of the Dutch Ministry of Education, Science and Culture. Because a code of conduct can only be useful if it reflects scientific practice, scientists as well as representatives of government and business were from the outset involved in developing this document.[23] Seminar participants were first shown the Code's 'basic principles', drawing the link back to the international ban on biological weapons:

23 Royal Netherlands Academy of Arts and Sciences 2008, *A code of conduct on biosecurity: Report by the Biosecurity Working Group*, available: <http://www.knaw.nl/publicaties/pdf/20071092.pdf> [viewed 20 January 2010].

> The aim of this code of conduct is to prevent life-sciences research or its application from directly or indirectly contributing to the development, production or stockpiling of biological weapons, as described in the Biological and Toxin Weapons Convention (BTWC), or to any other misuse of biological agents and toxins.

We then displayed the suggested elements of the Code with respect to 'research and publication policy' and 'accountability and oversight', in each case asking participants to tell us what they liked or disliked:

Research and publication policy

- Screen for possible dual-use aspects during the application and assessment procedure and during the execution of research projects.
- Weigh the anticipated results against the risks of the research if possible dual-use aspects are identified.
- Reduce the risk that the publication of the results of potential dual-use life-sciences research in scientific publications will unintentionally contribute to misuse of that knowledge.

Accountability and oversight

- Report any finding or suspicion of misuse of dual-use technology directly to the competent persons or commissions.
- Take whistleblowers seriously and ensure that they do not suffer any adverse effects from their actions.

The ensuing discussion invariably centred on what the application of such general principles would look like in practice. Seminar participants would typically pick out one or more of the above words and ask a question along the lines of 'What does that really mean?' What did it mean, for example, to 'screen' research? How and by whom are 'possible dual-use aspects' to be 'identified'? What about results that are not 'anticipated'? What does it mean to 'unintentionally contribute' to misuse of knowledge? What are good grounds for 'suspicion' of misuse of technology? What is the difference between a 'whistleblower' and a troublemaker? What constitutes taking something 'seriously'? Such questions, drilling down to the fine detail of how scientists' day-to-day working lives might be affected by a code of conduct, generated excellent and sometimes heated discussions among the seminar participants. Our role as facilitators, we reminded them, was not to answer their questions but to get an informed conversation started on possible solutions to the dual-use dilemma in the life sciences. At the end of each seminar, after thanking the participants for their contributions and providing our contact details, it was pleasing to hear conversation continuing in the corridors as people headed back to their places of work.

General Reflections

It was not the purpose of these seminars for the facilitators to advance a moral position on particular lines of research and publication decisions or to advocate any specific form of oversight to manage the dual-use dilemma. Rather, these were opportunities for loosely structured discussions moderated by impartial, non-government experts. Only rarely did a participant ask us directly for an answer to a question we posed, and such a request was always met with a deliberate and conspicuous sidestep. We were candid about not being able to agree among ourselves, and we admitted that the seminars themselves were an occasion to rethink our opinions, but beyond that we insisted on keeping a low profile. Each of us had come to this awareness-raising project with different intellectual backgrounds (virology, bioethics and political science), and it was sometimes difficult to reconcile our individual views on the kind of content to cover and the best process to follow. In preparing for and debriefing after each seminar, we saw the series evolve and improve. As facilitators, we reflected on what we judged to be mistakes in our own and each other's approach and sought to build on those aspects of previous seminars that seemed to work well.

At all times, however, we needed to be flexible and alert enough to facilitate conversation among seminar participants rather than try too hard to shape it. We needed to be careful, for example, not to be seen to be prefacing questions or pre-empting answers. Lest we appear to be pushing an agenda, we tried to ask open questions (for example, 'What is your moral position, if any, on X?') rather than leading ones (for example, 'Do you think X is immoral?'). Our language had to be as neutral as possible, abstract rather than personal, to maintain focus on the participants' own views (for example, 'One argument against that might be…', 'Another perspective you might come across is…', 'Can anyone think of a possible counterargument?'). In addition to being impartial, we needed to be inclusive. A consistent feature of all four seminars was that the more senior scientists in the room tended to speak first, at greater length, and contributed statements rather than questions. Most of their contributions were well informed and constructive, but we were sensitive also to the importance of ensuring that the more junior scientists and students had an opportunity to voice opinions as future research leaders. The aim overall was to draw people into articulating their reasoning on how they would navigate a dual-use dilemma, and to flush out the assumptions underlying particular points of view. Our facilitative tasks were to attend carefully to what was being said at the time, recall similar or countervailing points raised earlier, anticipate what might be raised next, and keep the conversation moving swiftly enough to cover all our themes.

The experience of designing and running the seminars was both challenging and rewarding, and it was unlike any other academic activity the facilitators had engaged in before. It was a privilege to discuss the dual-use dilemma with some of the most talented life scientists in Australia, and we made some useful professional contacts over the course of 2009. In Canberra, Melbourne, Sydney and Brisbane, we were greatly encouraged seeing seminar participants taking the time and effort to talk seriously about a vexing and intriguing issue. We were made to feel very welcome at every institution we visited, and each seminar proved a valuable learning experience for us too. Although limited in scope, our pilot seminar series seemed to confirm what anecdotal evidence had led us to suspect: that in Australia there is very little familiarity overall with policy issues surrounding the dual-use dilemma in the life sciences. The remedy we proposed, as simply and tactfully as we could, was to get people talking. However distasteful the topic of biological weapons might seem, it is nevertheless an increasingly important one for Australian life scientists and their colleagues around the world. We sought to raise awareness of the dual-use dilemma in a manner sensitive to but not beholden to the interests and concerns of life scientists. Scientific opinions are not the only ones that count; potentially all members of a given community have a stake in this issue. On the other hand, life scientists are historically not closely acquainted with the language and institutions of national security. As such, seminars such as ours might serve as a source of empowerment as well as information. By becoming more familiar with political and policymaking processes, life scientists might be able to suggest better ways of managing the security risks inherent in some research while minimising scientific opportunity costs.

In Australia and beyond, with the health and security stakes so high, there needs to be further exploration of education and awareness-raising as a long-term means of preventing the destructive use of science. Accordingly, the National Centre for Biosecurity is drafting a proposal that the Australian Government fund a nationwide Seminar Programme to raise awareness of the dual-use dilemma among Australian life scientists and students. The Programme would consist of 24 seminars (six in 2011, eight in 2012, and 10 in 2013) and three annual training days for seminar facilitators. The seminars would initially resemble those conducted in the 2009 pilot scheme, and updates and improvements would be made over the succeeding three years. Such an initiative would firstly support Australia's national security and public-health objectives. By raising awareness of biological-weapons risks in a manner sensitive to the interests of scientists, the seminars would facilitate conditions in which biological attacks are less likely to emanate from the Australian life-sciences community. Secondly, the Programme would support foreign-policy objectives by strengthening the fulfilment of Australia's international obligations under the 1972 BTWC. Specifically, Article IV obliges member states to take 'any necessary measures' to give effect to the Convention's prohibitions. Lastly, the initiative could provide members of the Australian life-sciences community with a respectable forum to articulate their interests and concerns regarding the dual-use dilemma.

Chapter 8: Bringing Biosecurity-related Concepts into the Curriculum: A US View

NANCY CONNELL AND BRENDAN MCCLUSKEY

The decades flanking the September/October 2001 terrorist incidents in the US (1990–2010) have seen a dramatic increase in concern with and attention to biological weapons (BW). The discovery of an extensive offensive BW programme in the former Soviet Union, the unsuccessful attempts of Aum Shinrikyo and US domestic terrorists to acquire, produce and disseminate 'weaponised' biological agents, and the anthrax attacks through the US Postal Service are among the events that have contributed to increased awareness of a possible biological threat.

While assessments of the actual threat remain controversial, the perceived threat already has led to extensive changes in the conduct and regulation of scientific activity in the US. In the past decade, concern with biological weapons and biodefence has been accompanied by massive increases in funding directed towards civilian biodefence: over $50 billion between 2001 and 2009.[1]

An enormous amount of federal effort and capacity is now directed towards select-agent research, in particular, and infectious-disease research in general.

Accompanying this push in infectious-disease research are requirements for compliance with increased regulatory activity at federal, state and institutional levels.[2] The changes include enhanced personnel and site-security oversight, consideration of delaying publication of relevant results, and greater regulation and management of experimental research. Thus far, systems of control have focused largely on laboratory biosafety and biosecurity, by regulating manipulation of and access to highly infectious organisms. The impetus for this has come largely from federal agencies.

1 Franco, C. 2009, 'Billions for biodefense: federal agency biodefense funding, FY2009–FY2010', *Biosecurity and Bioterrorism*, vol. 7, September, pp. 291–309.

2 Jaax, J. 2005, 'Administrative issues related to infectious-disease research in the age of bioterrorism', *Institute of Laboratory Animal Resources Journal*, vol. 46, pp. 8–14.

Also, a series of recent experiments in infectious-disease research (discussed in other chapters in this volume) have brought the concept of 'dual use' to the fore. For decades, the term 'dual use' was applied to the civilian/military duality. This concept has continued to evolve. Today the concern is that most technologies developed for legitimate purposes are intrinsically capable of being exploited for nefarious ones. It is not difficult to appreciate this potential in contemporary life sciences. There have been a number of pivotal technical advances in biomedical research over the past two decades. For example, the introduction of polymerase chain reaction in 1983 permitted the measurement of gene expression with previously unimaginable precision; the application has continued to develop novel applications. Current imaging techniques allow precise mapping of metabolic and signalling pathways, in real time and in whole animals, including humans. Nanotechnology and microfluidics have created more-effective delivery methods of drugs, hormones, and bioregulators. Increased information relating to physiology, behaviour and disease paves the way to new methods of controlling biological responses in medicine and improving human life. Yet, it can be argued that each of these advances is accompanied by the potential for malfeasance. In relation to dual-use concerns, the Fink and Lemon/Relman Reports[3] argued that the scientific community must increase its involvement in the development of policy. The creation of the National Security and Biosafety Board (NSABB) has been a useful exercise in focusing the attention of leaders in academic and commercial research on this topic.

Interest in dual use continues to grow. In 2009, the American Association for the Advancement of Science (AAAS) and the National Academies of Science (NAS) jointly published a report titled *A Survey of Attitudes and Actions on Dual-use Research in the Life Sciences*.[4] The results of the study suggested that the majority of life-sciences researchers in the US supported the concept of oversight models that rely on self-governance and responsible conduct, but that clarification of a number of issues is required. This included matters such as defining the scope of research and experiments of concern, establishing appropriate training mechanisms, and identifying ways that scientists can contribute to the prevention of misuse of scientific knowledge. These same issues were revisited in a series of recent workshops held by the AAAS to examine existing programmes in dual-use education and in biodefence policy training.[5] From these studies, it became clear that most academic institutions

---

3 National Research Council 2004, *Biotechnology research in an age of terrorism*, Washington, DC: National Academies Press; Institute of Medicine and National Research Council 2006, *Globalization, biosecurity and the future of the life sciences*, Washington, DC: NRC.

4 National Research Council/American Association for the Advancement of Science 2009, *A Survey of Attitudes and Actions on Dual Use Research in the Life Sciences: A Collaborative Effort of the National Research Council and the American Association for the Advancement of Science*, Washington, DC: NRC/AAAS.

5 American Association for the Advancement of Science 2009, *Building the Biodefense Policy Workforce*, Washington, DC: AAAS; American Association for the Advancement of Science 2008, *Professional and*

provide few resources for or demonstrate little interest in dual-use education. Further, educational materials are lacking, as are methods and analysis of their efficacy. The National Science Advisory Board for Biosecurity (NSABB) released its *Strategic Plan for Outreach and Education on Dual-use Research Issues* in 2008.[6] A joint letter[7] to the NSABB from the AAAS, the American Association of Medical Colleges (AAMC), the Association of American Universities (AAU), the Council on Government Relations (COGR), the Federation of American Societies for Experimental Biology (FASEB) and the Association of Public and Land-grant Universities (NASILGC) outlines these groups' apprehension with the mechanism of review, the determination of whether specific dual-use research would be categorised as being 'of concern', and the lack of clarity concerning liability issues that might lead to a dampening of scientific enterprise.

Dando[8] and others have called for the creation of a 'culture of responsibility'. This chapter will discuss approaches to this challenge in the current US academic environment. The idea of instilling a culture of social responsibility among scientists with respect to security issues is underpinned by the question of disclosure mechanisms, anonymity, and whistleblower protection. These are not novel topics, and are included in current standard biomedical-ethics curricula. However, disclosure of unusual or inappropriate activity takes on additional significance when the behaviour might be tied to national security.

Thus, in the last decade the US has witnessed the introduction of a number of new concepts to the life sciences. The process of doing science has been permeated by security and safety regulations that in turn have stimulated interest in the ethical and even moral issues related to the misuse of life-sciences research. Studies are accumulating to evaluate whether practising scientists are aware of these ideas, either by exposure or on their own, and which educational institutions have introduced these concepts into ethical-training programmes. Other chapters in this volume detail these studies in different parts of the world. Here, we discuss the challenge of introducing biosecurity

---

Graduate-Level Programs on Dual Use Research and Biosecurity for Scientists Working in the Biological Sciences, Washington, DC: NRC/AAAS.

6 NSABB 2007, *Proposed framework for the oversight of dual-use life sciences research: Strategies for minimising the potential misuse of research information*, Bethesda, MD: NSABB; NSABB 2008, Strategic plan for outreach and education on dual-use research issues, Bethesda, MD: NSABB.

7 Joint letter, 18 July 2008, to NSABB from the American Association for the Advancement of Science (AAAS), The American Association of Medical Colleges (AAMC), The Association of American Universities (AAU), The Council of Governmental Relations (COGR), The Federation of American Societies for Experimental Biology (FASEB) and The National Association of State Universities and Land-Grant Colleges (NASULGC, now the Association of Public and Land-grant Universities (APLU)), available: www.aau.edu/WorkArea/DownloadAsset.aspx?id=9740 [viewed 15 Mar 2010].

8 Atlas, R. and Dando, M. 2006, 'The dual-use dilemma for the life sciences: Perspectives, conundrums, and global solutions', *Biosecurity and Bioterrorism*, vol. 4, September, pp. 1–11; Revill, J. and Dando, M. 2008, 'Life scientists and the need for a culture of responsibility: After education...what?', *Science and Public Policy*, vol. 35, February, pp. 29–36.

and dual-use matters within the context of existing programmatic frameworks in a typical US academic biomedical-research institution. Over the past 15 years we have developed a number of avenues for introducing the concept of dual-use research to the university community at our institution. The first is through the federally mandated 'Responsible Conduct of Research' education of National Institute of Health (NIH)-sponsored trainees. The second route is via the Institutional Biosafety Committee, originally mandated by the NIH in the 1970s to review experiments involving recombinant DNA and since expanded to include infectious agents. The third avenue is the laboratory safety training mandated by the Occupational Safety and Health Association (OSHA) for all laboratory workers. The fourth route is through a robust biodefence 'certificate' academic curriculum, open to all students at the university regardless of programme (PhD, MS, MD, nursing, and so on). We propose a fifth approach using an institutionally based 'train-the-trainer' system of intercalating dual-use awareness into individual academic departments through periodic seminars and discussion groups. We discuss the strengths and limitations of each of these approaches in terms of topics, efficacy and audience.

Route One: Responsible Conduct of Research

An examination of the history of incorporation of ethical issues into the US curriculum will enrich this exploration of mechanisms for introducing biosecurity and dual use into the academic biomedical curriculum. Formalised ethics training was introduced just over two decades ago in the US federally supported scientific enterprise. The impetus was a series of fraud/misconduct cases at four research institutions in 1980 that were widely publicised, leading to widespread calls for a concerted effort to include ethics training within the medical school curriculum, originating from both lay and medical groups. The first congressional hearing uncovering additional cases took place that same year, in the Investigations and Oversight Subcommittee of the House Science and Technology Committee.

In 1985, Congress passed the Health Research Extension Act, which required that Health and Human Services (HHS) awardee institutions establish 'an administrative process to review reports of scientific fraud' and 'report to the Secretary any investigation of alleged scientific fraud which appears substantial'. The Final Rule, *Responsibilities of Awardee and Applicant Institutions for Dealing With and Reporting Possible Misconduct in Science*, was published in the Federal Register in 1989 and codified as 42 CFR Part 50, Subpart A. The Office of Research Integrity (ORI) was established in its current iteration — that is, independent of the funding agencies — in 1992. The Commission on Research Integrity published a report titled *Integrity and Misconduct in Research* in November 1995.[9] It contained 33 recommendations, among which was the requirement of funded institutions to establish educational programmes on the responsible conduct of research (RCR).

The term 'misconduct' has evolved from its original definition of 'fraud, fabrication and plagiarism' to include 'other serious deviations from commonly accepted practices'.[10] In 1999, policy was developed requiring all extramural research institutions to provide training in RCR to all staff who have 'direct and substantive involvement in proposing, performing, reviewing, or reporting research, or who receive research training, support by Public Health System (PHS) funds or who otherwise work on PHS-supported research projects even if the individual did not receive PHS support'. Eight topics are required in addition to misconduct (fraud, fabrication and plagiarism): data acquisition, sharing and management; conflict of interest; animal protection; human-subject protection; publication and authorship; mentor–trainee responsibilities; peer review; and collaborative science. Scientific research is conducted in a constantly changing environment and RCR training has undergone gradual shifts in focus. Regulatory changes, electronic publishing, and data sharing have compelled adjustments or additions to the topics. The policy was suspended in February 2001 pending review and, interestingly, a ruling on whether the document should have been issued as a ruling remains suspended. Whistleblower protection was also reviewed, although the final rule has been pending since January 2001. However, the Whistleblower Protection Act of 1989 protects federal employees, individual institutions and corporations have their own protection policies implemented at state and institutional levels.

It would appear from this brief review of the history of RCR training and guidance in the NIH that this programme would be an excellent framework for the introduction of dual-use concepts to life scientists. Indeed, the Office of Intramural Research at the NIH has already explored new case studies and scenarios ('Science and Social Responsibility — Dual Use Research 2009' [11]) for inclusion in RCR training within the NIH's own intramural programme, which requires annual ethics training for all regular NIH employees — not only trainees.

Whether the RCR mechanism will provide adequate training of dual-use issues remains an important question. Recent studies of standard RCR training methods

9 Rhoades, L. J. 2004, *New Institutional Research Misconduct Activity: 1992–2001. Office of Research Integrity*, available: ori.dhhs.gov/education/products/rcr_misconduct.shtml [viewed 15 Mar 2010].

10 American Association for the Advancement of Science and the US Office of Research Integrity 2000, *The Role and Activities of Scientific Societies in Promoting Research Integrity. A Report of a Conference*, available: http://www.aaas.org/spp/sfrl/projects/report.pdf.

11 NIH Committee on the Conduct of Science 2009, *Science and Social Responsibility — Dual Use Research 2009*, available: www1.od.nih.gov/oir/sourcebook/ResEthicsCases/2009cases.pdf [viewed 15 Mar 2010].

have pointed to wide variation in both approaches and efficacy. Antes *et al.*,[12] who concluded that effectiveness was 'modest', carried out a meta-analysis of ethics instruction in the sciences. They noted that success was tied to course structure (case-based illustration and discussion was more effective than lecture) and context (instruction separated from standard curricula rather than included within existing courses). Others[13] argue that all trainees in our universities should be expected to understand basic principles of academic integrity and, further, to gain expertise in ethical issues in their individual fields. The study of Heitman *et al.*[14] observed a disheartening lack of knowledge among trainees upon entering graduate school, irrespective of previous research experience, ethics training, or country of origin; the authors suggest RCR training might be modified to adjust to gaps in knowledge and experience. Finally, a troublesome study by Anderson *et al.* examined early- and mid-career NIH-funded scientists who had received NIH-mandated RCR training.[15] Not only had many of the respondents little to no recollection of that teaching, but the study also found under some conditions a positive correlation between research-integrity training and behaviour that was inconsistent with that teaching. Critics of the entire RCR training enterprise claim that scientists as educated adults already have a moral framework within which the core concepts of RCR are adequately contained. These and many other studies suggest that our academic institutions should consider alternative educational methods for effective ethics training; and dual-use awareness should be included in the discussions.

In a December 2009 editorial titled *Bringing a 'Culture of Responsibility' to Life Scientists*,[16] Malcolm Dando pointed out that many researchers consider RCR training adequate for developing a culture of responsibility. Dando further observed that neither the ORI nor professional societies with similar agendas, such as the NAS or the Royal Society, had yet incorporated dual-use issues in any formal way. Washington's recent 'National Strategy for Countering Biological Threats'[17] contains as its second objective the 'reinforce[ment of] norms of safe

---

12 Antes, A., Wang, X., Mumford, M. D., Brown, R. P., Connelly, S. and Devenport, L. D. 2010, 'Evaluating the effects that existing instruction on responsible conduct of research has on ethical decision making', *Academic Medicine*, vol. 85, March, pp. 519–26.

13 Bulger, R. E. and Heitman, E. 2007, 'Expanding responsible conduct of research instruction across the university', *Academic Medicine*, vol. 82, September, pp. 876–8.

14 Heitman, E., Olsen, C. H., Anedtidou, L. and Bulger, R. E. 2007, 'New graduate students' baseline knowledge of the responsible conduct of research', *Academic Medicine*, vol. 82, September, pp. 838–45.

15 Anderson, M. S., Horn, A., Risbey, K. R., Ronning, E. A., De Vries, R. and Martinson, B. C. 2007, 'What do mentoring and training in the responsible conduct of research have to do with scientists' misbehavior? Findings from a national survey of NIH-funded scientists', *Academic Medicine*, vol. 82, September, pp. 853–60.

16 Dando, M. 2009, 'Bringing a "culture of responsibility" to life scientists', *Bulletin of the Atomic Scientists*, 18 December, available: http://www.thebulletin.org/web-edition/columnists/malcolm-dando/bringing-culture-of-responsibility-to-life-scientists [viewed 15 March 2010].

17 National Security Council 2009, *National strategy for countering biological threats*, available: http://www.whitehouse.gov/the-press-office/president-obama-releases-national-strategy-countering-biological-threats [viewed 15 March 2010.

and responsible conduct' by developing appropriate training programmes and materials. Dando mused whether these strategies will be implemented in time for the Seventh Review Conference of the BWTC in 2011.

Integration into Ethics and Responsibility Training: A Case Study

What follows is a description of the gradual incorporation of biosecurity and dual-use issues into the RCR curriculum of the University of Medicine and Dentistry of New Jersey (UMDNJ) that began in 1994 at the Newark branch of the Graduate School of Biomedical Sciences (GSBS). The GSBS in Newark presents its RCR course for PhD students at the end of their second year, just as they finish the didactic segment of their training and enter the laboratory full time. The course is team-taught and the lecturers represent various departments and regulatory cores of the institution. The teaching style is a mix of lecture and interactive case-study discussion. In 1994, a single lecture, titled 'Biological and Toxin Weapons', was introduced into this course and focused on the history of biological and toxin weapons use, the nature of the agents and the difficulties in working with them, the past offensive programmes of the US, UK, Japan and USSR, weapons-testing programmes, and the history and development of the BTWC. Discussion topics included the verification protocol that was under development and the responsibility of scientists to recognise and support the BTWC. Students were urged to think about problems in detection of production or weaponisation methodology and learned about the pledge, circulated by the US Council for Responsible Genetics in 1989, that scientists not participate knowingly 'in research and teaching that will further the development of chemical and biological agents'. Although dual-use issues were not yet a primary focal point of the BTWC Review Conferences, codes and the dual-use dilemma were already part of the discussion in many scientific circles.

There were two subsequent changes in the focus of UMDNJ's lecture as the years went by. One was in 1999 when the institution established a biodefence-research programme, forming the UMDNJ Center for BioDefense, accompanied by construction of a new Biosafety Level Three laboratory for the study of infectious respiratory micro-organisms, including select agents. The RCR lecture in bioweapons expanded at this point to include the topics of biosafety and biosecurity, natural versus man-made outbreaks of disease, and so on. The Center for BioDefense had a strong Emergency Response training component that further expanded the scope of the lecture. The BTWC and the responsibility of scientists in maintaining awareness of a possible biological arms race, including the UN inspection teams, remained the cornerstone of the lecture. The second major change in structure of the RCR lecture was in Spring 2002, four months after the anthrax attacks. At this juncture, the lecture began to include yearly updates in the anthrax mailing case, biological terrorism, and so on. In fact, a group of graduate students approached the Center for BioDefense asking for full-length courses in both basic science and policy. In response to this request, a certificate programme in biodefence was developed (discussed below).

UMDNJ is a large institution, with three branches of the graduate programme in different parts of the state of New Jersey. Inquiries directed at the Center for BioDefense from other segments of the GSBS suggested that students across the university would benefit from knowledge of such things as select-agent research, and biosafety and biosecurity regulations. Therefore, approximately 75 students per year are taken through a two-hour lecture and interactive discussion of the history of biological weapons, arms control, codes of conduct and the dual-use dilemma.

Nevertheless, using the RCR to introduce biosecurity and dual use-issues has a number of limitations. For example, only graduate students take this course. How would Principal Investigators (PIs) be included in the programme? The NIH and a very small number of universities have included PIs in their training programmes, but this is rare.21 The RCR requirement for trainees was initiated in 1995, and assuming these first students left college soon after, they should now be at assistant- or associate-professor level, or the equivalent in industrial settings. These researchers have been surveyed regarding the effectiveness of RCR training, as discussed above.[18]

There are other groups of scientists who play significant roles in the scientific enterprise who are often not included in RCR training. The first comprises those in post-doctoral training: recently, the National Postdoctoral Association has introduced materials for RCR training on its website[19] and the NIH now requires RCR training for recipients of its 'K-series' of awards, for which post-docs and junior faculty are eligible. The second group are the research technical staff: it is possible that these groups can be reached through laboratory safety training, described below.

---

18 Anderson *et al*. 2007, op. cit.; Antes *et al*. 2010, op. cit.

19 National Postdoctoral Association 2009, *Tailoring RCR programs for postdocs*, available: http://www.nationalpostdoc.org/publications/rcr/112-pda-toolkit-tailor-to-postdocs [viewed 15 March 2010].

Route Two: The Institutional Biosafety Committee

A second introductory route of biosecurity and dual use into the curriculum is through the Institutional Biosafety Committee (IBC). IBCs were established in the 1970s in response to alarm and concern in the scientific community over the potential dangers of the then novel recombinant DNA technology. The NIH *Guidelines for Research Involving Recombinant DNA Molecules*[20] have been continually updated and are now under the charge of the Office of Biotechnology Activities. The Recombinant DNA Advisory Committee (RAC) members are responsible for oversight of these activities by interpreting the NIH Guidelines (latest version, September 2009). Appendix G is the section that deals with physical containment and biosafety. In June 2009, a 'Tool for the Self-Assessment of the Institutional Biosafety Committee and Programme of Oversight of Recombinant DNA Research'[21] was released, which allows individual IBCs to evaluate their effectiveness and compliance with federal regulations. The IBC reviews and approves all research involving 'non-exempt' recombinant DNA, pathogenic micro-organisms and/or potentially infectious materials requiring work at Biological Safety Level 2 (BSL-2) or above. Research protocols are prepared by principal investigators and submitted for appraisal before the work is begun: the major review focus is the safety of workers carrying out the experiments and the community, both within and outside the institution.

As the purview of the IBCs has expanded from recombinant DNA to include pathogen research, these committees have been identified as a control point for oversight of research with dual-use potential. The Fink Report advocated expanding the responsibilities of IBCs to include biosecurity and dual-use concerns.[22] However, this suggestion has been met with criticism: in the NAS report *Science and Security is a Post-9/11 World,*[23] David Relman is quoted as saying, 'Today's IBC's can't do biosecurity because the members have not been adequately informed about how you think [about] biosecurity, how you think about the potential misuse of science'.[24] While Relman's point is well made, the

---

20 Office of Biotechnology Activities 2002, *NIH Guidelines for Research Involving Recombinant DNA Molecules*, available: oba.od.nih.gov/rdna/nih_guidelines_oba.html [viewed 15 March 2010].

21 Office of Biotechnology Activity 2009, *Tool for the Self-Assessment of the Institutional Biosafety Committee and Program of Oversight of Recombinant DNA Research*, available: oba.od.nih.gov/rdna_ibc/ibc.html [viewed 15 March 2010].

22 National Research Council 2004, *Biotechnology research in an age of terrorism*, Washington, DC: National Academies Press.

23 National Research Council 2007, *Science and Security in a Post 9/11 World: A Report Based on Regional Discussions Between the Science and Security Communities*, Washington, DC: National Academies Press.

24 Relman, D. 2006, Remarks made at the Committee on a New Government–University Partnership for Science and Security Western Regional Meeting at Stanford University, 27 September, available: www7.

past few years have seen a wealth of information and scholarly articles addressing these issues. Several training modules have been developed for online use and IBC members might be required to undergo these training modules. Much of this dual-use material was analysed at a series of workshops held by AAAS[25] and reviewed by the NSABB[26]. For example, online modules are sponsored by the following organisations:

- Duke University (SERCEB) at: www.sercebtraining.duhs.duke.edu/
- Federation of American Sciences at: www.fas.org/biosecurity/education/dualuse/index.html
- NIH Office of Research Integrity at: www1.od.nih.gov/oir/sourcebook/ResEthicsCases/2009cases.pdf
- The Center for Arms Control and Nonproliferation at: www.politicsandthelifesciences.org/Biosecurity_course_folder/base.html.

Other contributors to this volume have also outlined additional resources.

Integration into the Institutional Biosafety Committees: A Case Study

The IBC at UMDNJ in Newark has taken specific steps to begin introducing dual use into its agenda. Members of the IBC have all taken the online dual-use-awareness modules developed by Duke University (SERCEB) and they now evaluate submitted protocols for dual-use potential in their discussions. The Newark IBC is currently considering appropriate language to incorporate questions regarding dual use into the IBC protocol application itself, thereby involving the PI directly. The Department of Environmental and Occupational Health and Safety Services has published the first of a series of articles in its monthly newsletter on dual-use experimentation. This newsletter is widely distributed across the entire campus. The goal is to introduce these ideas across the university community, reaching PIs, trainees and staff in all fields, regardless of whether they work with recombinant DNA or pathogenic organisms.

As with integrating dual-use education into the RCR component of life-sciences instruction, there are drawbacks to the approach of solely relying on the IBC as a vehicle for dual-use education. In addition to the possible lack of expertise and experience discussed above, not all research with potential dual-use application is captured by the IBC as it is currently configured. It focuses on experiments using recombinant DNA and/or highly infectious agents. Research

nationalacademies.org/stl/202006.pdf [viewed 15 March 2010].

25 American Association for the Advancement of Science 2009, *Building the Biodefense Policy Workforce*, Washington, DC: AAAS.

26 NSABB 2008, op. cit.

not involving these activities will not be captured. Most, but not all, life-sciences research uses molecular biology and cloning, but protocols from fields using technologies identified as dual use in nature, such as neural imaging or nanotechnology — projects that do not use rDNA — will not be reviewed. Other regulatory committees in biomedical research are those that oversee laboratory-animal welfare and human-subjects protection: these institutional committees might be engaged to evaluate proposals for dual-use potential.

These limitations argue for a broader approach, one that includes all kinds of research and targets executives, administrators, PIs, technical staff and trainees. Laboratory Safety Training is a requirement for all laboratory workers, and provides a third level of introduction to dual-use issues.

Route Three: Laboratory Safety Training

A third route of entry is through the Laboratory Safety Training required for all laboratory workers: PIs, post-doctoral fellows, graduate students, technicians and other staff. The OSHA has identified within its array of standards for general industry those with specific application to laboratories.[27] Topics include chemical safety/'right-to-know',[28] hazardous-waste and regulated medical-waste handling, fire safety, personal protective equipment, and emergency procedures. The training at UMDNJ lasts two hours and is given at the time of hire, followed by a short refresher course every other year thereafter. Laboratory biosafety and biosecurity are topics usually covered and dual use might be incorporated. However, we have found the brevity of such an introduction to a complex issue like dual-use research is inadequate to the task of successfully increasing awareness. Indeed, even the current methodology used for training and education in the responsible conduct of research — including the concept of dual use — have come under increased scrutiny and criticism, as discussed above.

Route Four: The Biodefense Certificate Programme

Faculty of the Center for BioDefense developed a certificate programme in biodefence for its PhD, MD and MS students. The programme comprises five

---

27 United States Department of Labor, Occupational Safety & Health Administration 1970, *Occupational Safety and Health Act of 1970*, available: http://www.osha.gov/SLTC/laboratories/standards.html.

28 Environmental Protection Agency 1986, *Emergency Planning and Community Right-to-Know Act of 1986*. available: http://frwebgate.access.gpo.gov/cgi-bin/usc.cgi?ACTION=BROWSE&TITLE=42USCC116.

compulsory courses and one elective. Of those required are two fundamental biomedical science (specified by the student's degree programme) and three biodefence-specific courses. The first of the latter is a weapons-survey course: biological and toxin agents are reviewed with respect to virulence, pathogenesis, route of infection/intoxication, treatment, history and potential use. The second focuses on the molecular biology of select agents (bacteria and viruses) and focuses on key papers, both classic and in current literature. The students read and analyse the science and policy implications of studies such as the three iconic papers describing (1) the mousepox IL-4 virus,[29] (2) the 1981 flu reconstruction,[30] and (3) the botulinum contamination of the milk supply.[31] The third is a seminar on contemporary topics, in which students are asked to prepare and discuss current issues in biodefence research, policy and history, including ethics, the responsibility of scientists as citizens, and so on. These courses contain extensive sections on dual-use analysis of contemporary topics in biodefence research. The primary question here is whether this thorough examination of the issues would be impractical for all researchers, either during or after PhD/MS awards. In the current climate of steadily increasing regulatory and compliance requirements, the answer would likely be 'yes'.

Route Five and the Challenges Ahead

In the current climate made competitive by limited funding, and in the absence of a federal mandate, we find concerns among many of the faculty over the possible introduction of dual-use awareness training, consistent with points made by the joint letter to the NSABB from a coalition of scientific societies discussed previously.[32] Many researchers feel they have little time to spend on anything other than the overwhelming demands of staying competitive in contemporary research. A common complaint is that there are many regulatory and compliance requirements in place that encumber the effective progress of science; adding another layer of regulation will be met with dismay by many scientists. Others are troubled by what they consider insufficient guidance on both the definition of 'dual use' and the consequences of a positive identification. Exactly what is dual-use research, and, if one's work is determined to be of dual-use 'concern', what then? There are fears that there is not enough expertise to

29 Jackson, R. J., Ramsay, A. J., Christensen, C. D., Beaton, S., Hull, D. F. and Ramshaw, I. A. 'Expression of mouse interleukin-4 by a recombinant ectromelia virus suppresses cytolytic lymphocyte responses and overcomes genetic resistance to mousepox', *Journal of Virology*, vol. 75, pp. 1205–10.

30 Perrone, L. A. S. and Tumpey, T. A. 2007, 'Reconstruction of the 1918 pandemic influenza virus: how revealing the molecular secrets of the virus responsible for the worst pandemic in recorded history can guide our response to future influenza pandemics', *Infectious Disorders Drug Targets*, vol. 7, pp. 294–302.

31 Wein, L. M. and Liu, Y. 2005, 'Analyzing a bioterror attack on the food supply: the case of botulinum toxin in milk', *Proceedings of the National Academies of Sciences*, vol. 201, pp. 9737–8.

32 Joint letter, 18 July 2008 (see note 7).

anticipate the possible dangers of certain experiments; some think the Australia group should have been able to predict that the IL-4 recombinant ectromelia virus would have a lethal phenotype.[33]

There is a logical connection between dual-use awareness and the generation of a 'pledge' or 'code of conduct'.[34] Although this is outside the scope of the present discussion, we have found that many scientists who think about the dual-use dilemma and an associated code of conduct inevitably arrive at the issue of whistleblowing and whistleblower protection. Might the development of a 'Culture or Responsibility' lead to a culture of accusation and suspicion? A related issue that arises frequently is that of liability: if an entity recognises an experiment or line of inquiry as being of dual-use potential, what is the responsibility of the funding agency or institution sponsoring the research in the event that the information or reagent does lead to a biocrime, a terrorist incident or even a catastrophic event? Indeed, these critical questions were raised in response to the NSABB's 2008 *Strategic Plan*.[35]

Despite these concerns, we have detected a gradual thaw in attitudes toward dual use over the past decade. The RCR course discussion of dual-use experiments of concern has been received eagerly across the university's several campuses. The IBC of UMDNJ in Newark is thinking energetically about ways to incorporate dual-use issues into our review. Some of our colleagues are finally willing to include dual-use questions embedded within larger exam questions in immunology and infectious-disease courses. The University's Department of Environmental and Occupational Health and Safety Services — an arm of the administrative branch — has embraced the issue and recognised the importance of disseminating information and resources. Our next step will be to recruit interested faculty in each department and begin to introduce seminars on dual use as regular yearly or twice-yearly events. We will focus on faculty who are already committed to teaching and mentoring. Our feeling is that discussions can be introduced at many different levels across the institution, creating a 'web of instruction'. Gradually, an appreciation of the complexity of the dual-use dilemma will become part of the scientific idiom.

---

33 Muellbacher, A. and Lobigs, M. 2001, 'Creation of killer poxvirus could have been predicted', *Journal of Virology*, vol. 75, pp. 8353–55.

34 Atlas and Dando 2006, op. cit.

35 Joint letter, 18 July 2008 (see note 7).

PART 3

THE WAYS FORWARD

Chapter 9: Implementing and Measuring the Efficacy of Biosecurity and Dual-use Education

JAMES REVILL AND GIULIO MANCINI

Teaching security-related issues to science students is increasingly salient in the international security discourse, yet despite the calls for greater education, research conducted by the authors and others contributing to this volume (as in the chapters by Friedman and Minehata and Shinomiya) demonstrates that the calls for, and interest in, biosecurity education have too infrequently been converted to commensurate activity at the level of the practising life scientist. Indeed, it remains clear from previous chapters that there are currently limits to the extent of biosecurity education and the process of promulgation and implementation of education has been slow, something stymied by practical, ethical and philosophical considerations which need to be addressed in the process of moving from aspirational declarations by states and other organisations to concrete action (see the chapter by Johnson).

The purpose of this chapter is to posit some partial responses to the questions posed in the introduction, specifically questions related to the 'Who, What and How' of education. These will be based on the authors' experiences derived from a joint project between the Landau Network Centro Volta (LNCV) and Bradford Disarmament Research Centre (BDRC) on sustainable biosecurity and dual-use education. This project represents one experimental applied approach to the implementation and promulgation of a specific element of biosecurity education that focuses on the 'wider ethical/arms-control issues' mentioned earlier by Rappert. In this regard, biosecurity in the context of this chapter relates less to the notion of laboratory biosecurity. Rather, is much more closely linked to concerns over the limited awareness amongst scientist both of the existence of measures intended to prevent and prohibit the malign exploitation of the life sciences, and how benignly intended research could be misapplied.

This chapter firstly provides a brief elaboration of the phases of activity in the LNCV-BDRC project, beginning with a gap analysis of university curricula in Europe and proceeding with the construction of a collaborative

Biosecurity Education Network. Secondly, the chapter outlines the process of implementation testing using tailored material that was presented primarily to life-science students under the supervision of network members working within a number of departments in European universities. Thirdly, this chapter provides information on some of the lessons learned over the course of implementation tests, before underlining the vital importance of other complementary intervention points which could be addressed in the process of moving from aspirational statement of interest to concrete activity.

Biosecurity Education Survey

The joint LNCV-BDRC project was implemented with the intent of firstly developing an understanding of the extent of, and attitudes to, education on wider ethical/arms-control and dual-use issues within life-science degree courses. Also underlying this objective was an aim to identify means and methods through which biosecurity education could be more effectively promulgated across the life-science community in Europe. In this sense, the LNCV–BDRC project was initiated with a focus on university students. Whilst recognising that other targets could be considered for inclusion, engaging students was deemed particularly important because the development of advanced biological weapons is likely to require some degree of sophisticated scientific capability. This logically points advocates of education towards a primary target group consisting of individuals trained in the life sciences to at least undergraduate level. Moreover, engagement with this sector could be particularly beneficial in the long term in feeding into a sustainable culture of responsibility built by the next generation of life scientists.

In the first phase of the LNCV–BDRC project, the aim was to produce a gap analysis which assessed the extent of, and attitudes to, biosecurity-related content in universities across Europe using a sample of 142 degree courses (undergraduate and master's level) at 57 universities in 29 countries. [1] This phase involved an assessment of the existence and extent of biosecurity-related content through the development of sample and search terms and the implementation of an investigation into available content (including online programmes of study and syllabi); but also the identification of key individuals within departments with whom we subsequently confirmed the provisional results and elicited an understanding of attitudes through follow-up correspondence.

---

1 The Member States of the European Union but also Norway and Switzerland. The sample was selected on the basis of a balance between international ranking and geographical representation, see Mancini, G. and Revill, J. 2008, 'Fostering the Biosecurity Norm: Biosecurity Education for the Next Generation of Life Scientists', Working paper produced for the Landau Network-Centro Volta, available:http://www.centrovolta.it/landau/2008/11/20/FosteringTheBiosecurityNormBiosecurityEducationForTheNextGenerationOfLifeScientists.aspx.

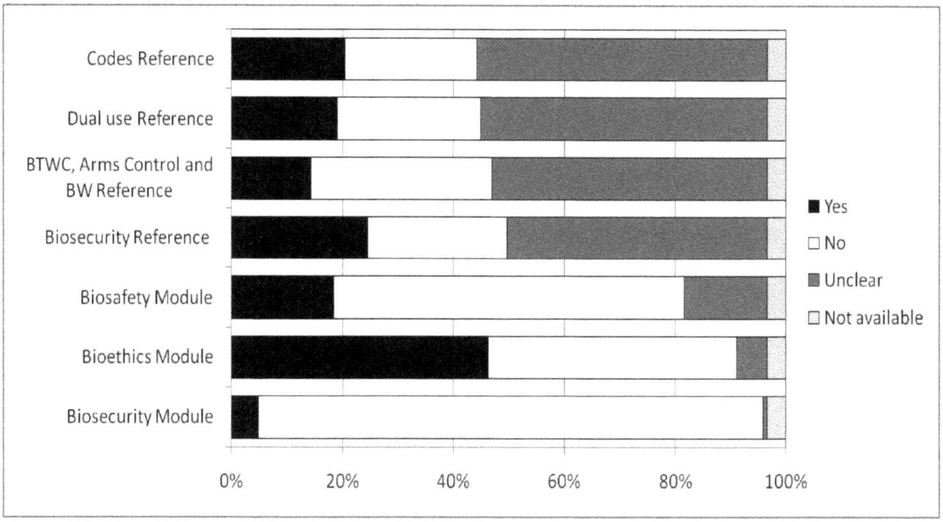

Figure 11: Overview of results from investigation into the biosecurity contents of European life-science and biotechnology courses

This investigation was not intended to produce statistically significant results and suffered a number of problems in determining content with certainty, particularly vis-à-vis 'references' to concepts related to biosecurity. Indeed, what constitutes a 'reference' ranged from one or more dedicated lectures, to a reference during a lecture. Nevertheless, the process provided an illustrative snapshot of the situation in Europe with regards to biosecurity and related components. We found a total of 37 (26 per cent) life-science and biotechnology degree courses from our sample where there was clear evidence of a reference to biosecurity; 22 (15 per cent) courses mentioned biological weapons, the BTWC or other arms-control agreements; in 29 (20 per cent) courses there was clear reference to the dual-use issue and finally, in 31 (22 per cent) courses there was a reference to codes of conduct, practice or ethics.

In terms of generating an understanding of attitudes to biosecurity education, responses to questions regarding whether biosecurity was perceived as important largely fell into three categories, with the majority falling in the first two. In the first category were a number of examples where curricula clearly didn't offer a specific module in their syllabi, or respondents felt the topic was irrelevant. Thus, for example, it was suggested 'we do not teach anything to do with the BTWC. I'm not sure if teaching such material on the BTWC would be helpful to our students unless they went into the field'. In the second category were those that felt the topics were relevant but were constrained by practical factors. Therefore, for example, one participant noted 'the main reason [for not having a specific biosecurity module] being the limited time we have to expose our students to science & society issues'. Finally, there were those that were clearly interested in the inclusion of this material as a topic of discussion; for example, one participant stated, 'I will take profit of your enquiry and integrate these issues in the introductory course starting with this year'.

In addition to the collection of data to determine the extent of biosecurity education in Europe, the process was useful in constructing the collaborative network through the identification of, and engagement with, individuals from various disciplines that were interested in advancing education on these subjects. This process was to some extent confirmed through two workshops held in Como in 2008 and 2009 titled 'Fostering the Biosecurity Norm: An Educational Module for Life-science Students' and 'Biosecurity, Biosafety and Dual-use Risks. Trends, Challenges and Innovative Solutions', respectively, which brought together life-science educators and security experts to discuss these topics. The process of conducting a gap analysis had a clear secondary value in building a biosecurity education network and providing a means through which the target audience could be reached and their attention and active engagement secured.

Building a Biosecurity Education Network

Indeed, the development of an informal collaborative network of professional actors interested in biosecurity or dual-use education has proved a useful model through which the process of framing and implementing education can be advanced taking into consideration input from a range of actors. As such, this remains one possible practical response to the question posed by Rappert on how audiences and practising scientists can be reached. Camarinha-Matos and Afsarmanesh have defined a collaborative network as being 'constituted by a variety of entities (for example, organisations and people) that are largely autonomous, geographically distributed, and heterogeneous in terms of their operating environment, culture, social capital, and goals. Nevertheless, these entities collaborate to better achieve common or compatible goals.' [2]

Such a model is certainly consistent with the collaborative network established in the LNCV–BDRC project, which consists of a number of independent academic departments and research organisations covering a range of disciplines (life sciences, biotechnology, bioethics, law, political science, security studies and international relations) in various countries in Europe (including Sweden, Italy, Spain, the Netherlands, the UK, Portugal and Germany), and which have in a variety of ways provided input from different perspectives that cumulatively contributed to the advancement of the network and progress in the objectives.

2 Camarinha-Matos, L. M. and Afsarmanesh, H. 2004, 'Collaborative networks: a new scientific discipline', *Journal of Intelligent Manufacturing*, vol. 16, 2005, available: http://www.uninova.pt/~cam/Papers/ColNetsJIM.pdf.

In this regard, the collaborative network can be seen as facilitating both the process of building bridges between different communities and academic disciplines and enhancing the quality of material and the implementation of lectures. Contributions to the legal aspect of biosecurity education have been made, for example, through engagement with the Institute for International Legal Studies at the National Research Council (ISGI-CNR) in Italy and the Department of Penal Law at the University of Granada in Spain. Cooperation with legal experts was particularly fruitful and led to the distillation of domestic legal references developed by a number of countries into accessible material more readily usable by science students and lecturers. More ethically and philosophically orientated consideration has also been advanced through engagement with the Department of Philosophy at Delft University of Technology in the Netherlands. Discussions with scientists and life-science lecturers within the network, such as those in the National Council of Biologists of Italian Universities, have been particularly useful for enhancing the quality and implementation of material. Firstly, they have provided feedback and advice on the optimisation of both teaching material and the delivery of lecturers to specific audiences; and, secondly, they have advised us on how implementation challenges can best be overcome.

Implementation Tests

Using nodes within the network developed through the survey and the Educational Module Resource elaborated upon by Whitby and Dando in another chapter in this volume, the LNCV–BDRC project has begun the process of testing material in several universities across Europe. This process was facilitated by contacts in the network who agreed to promote lectures or seminars organised by local professors, using agreed elements of material (summarised below in Table 3), as the basis for implementation tests tailored to their local contexts.

In some cases, implementation tests were conducted independently by life-science educators who developed their own lectures from material developed for the EMR and translated this into the local language where necessary. This suggests that in some cases the implementation of this approach to biosecurity education may merely require directing individuals to structured and accessible educational materials that can be used as a solid base for lectures which necessarily move beyond a scientific focus into broader social, economic and ethical territory. In other cases, network members began a process of engagement with the authors in which components of lecture courses were negotiated. In yet other cases, LNCV–BDRC project officers conducted lectures that were agreed with professors and tailored in time, content and delivery to suit the local audience.

Table 3: Summary of material in the EMR

Biological Weapons, Biowarfare and Bioterror • Biological Warfare from Antiquity to World War II • Biological Warfare in World War II • Biological Warfare During the Cold War • Assimilation of BW in Military Programmes and Calculation of Effects • Bioterrorism	**The Dual-use Risks and the Responsibilities of Scientists** • BW, Biosecurity and Ethics • Possible Hostile Applications of New Technologies • Hostile Application of Biology is Prohibited • Dual-Use: The Fink Report • Dual-Use: Examples • The Lemon-Relman Report • Weapons Targeted at the Nervous System • Possible Regulations of the Life Sciences?
The International Prohibition Regime and Implementation • International Legal Agreements • Development of the BTWC: 1980-2008 • The First Intercessional Process: the 2005 BTWC Meetings on Codes of Conduct • The BTWC Second Intercessional Process 2007–11 • International Customary Law • United Nations Security Council Resolution 1540 • EU Legislation and Directives on Export Controls • National Legislation	**The Web of Prevention** • Building an Effective Web of Prevention

In all cases there was not sufficient time to cover every component part of the EMR (which is intended as a comprehensive resource), nor was it necessarily important for undergraduate students in different scientific disciplines of relevance to learn all the information outlined in Table 3. However, it was important to ensure that the key elements were covered. In some cases this required a degree of discussion and content negotiation with network members where implementation tests were taking place. This was particularly important for engagement with individuals with scientific backgrounds who were more interested in focusing on examples of dual-use research.

This issue is significant in relation to the question posed in the introduction as to how to secure audience attention and active engagement. Three key areas are noteworthy in this regard. Firstly, it is apparent that an extensive section on the history of offensive biological weapons was perceived as unwelcome and unnecessary. Such a response is understandable in courses focused on science. However, the presentation of specific case studies can provide a stimulating discussion, but also an overview of the history of offensive biological-weapons programmes, which remains important in dispelling the myths amongst some in the scientific community that biological weapons have either not been used in warfare, or have been the preserve of other, 'bad' scientists. Secondly, some commentators were reluctant to include significant discussion on either arms-control measures or legislation on the grounds that this may be irrelevant, boring, or incomprehensible to science students. Whilst recognising such concerns, it is contended that a basic understanding of the legal and regulatory measures affecting scientists remains an essential component of biosecurity education because they can have a direct effect on their future activities.

Finally, in addition to the shortening of material to comply with the requirements of lecturers, it is apparent there is a need to include material which explicitly acknowledges the positive aspects of science, in order to both avoid the project being seen as in any way 'anti-scientific' and to avoid 'scaring' students away from science. On this basis, in more recent implementation tests we have been very clear that any security concerns need to be kept in perspective and the vast array of positive benefits of science in responding to societal challenges in the sphere of, *inter alia*, health, energy, food security and economic development are acknowledged at the beginning of the lectures. Indeed, for many in the scientific community, this is an essential component to avoid dissuading or deterring individuals away from the life sciences and prevent the appearance of being somehow anti-scientific.

Implementation Test Results

As part of the implementation test process we also attempted to evaluate the success of our activities. In the short term, there are a number of means of measuring the success of a lecture in raising awareness. These include pre- and post-lecture surveys (which could simply be done through raising hands), examination questions and seminar discussion-group demonstrations of understanding. In this project we opted to primarily use post-lecture surveys.

Using this approach it has been possible to determine the success of an individual lecture and, indeed, the results from implementation tests conducted with network members. These appear positive, with a significant percentage of participants indicating that they thought their knowledge of biosecurity and other related topics had been improved, as illustrated in Figure 12.

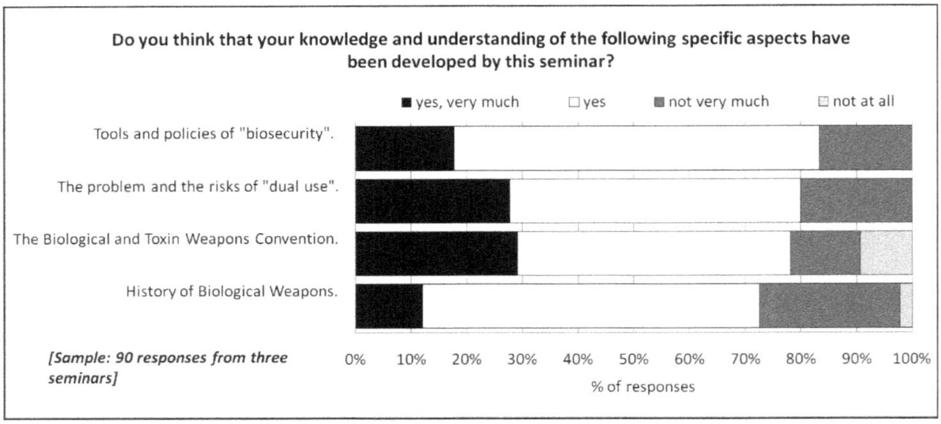

Figure 12: Responses to post-lecture questionnaire

It was also apparent that a significant majority of participants felt the lectures were both interesting and important, with 99 per cent appreciating the seminars and agreeing that raising awareness of biosecurity-related issues should be promoted among students, something evident in Figure 13.

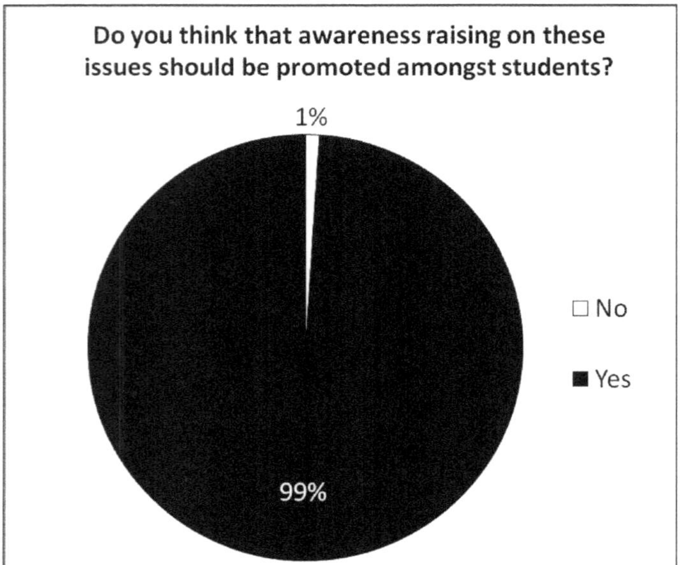

Figure 13: Responses to post-lecture questionnaire II

Such support for the promotion of biosecurity education is important to feed into a second test of success, that of sustainability. Sustainability is important in meeting the objective of building a culture of awareness which passes material on to subsequent generations and there are a number of metrics that can be used individually or in combination to test sustainability effectively. Examples include the following:

- post-lecture discussions with staff and students
- follow-up questionnaires
- requests for information on subsequent activity
- monitoring of website providing educational material related to biosecurity to determine the frequency and geographic location of downloaded material.

Based on these criteria, we can claim a degree of success in that network contacts in several cases have intimated that, given the interest raised and the available reference materials, they intend to both recommend and replicate lectures in future academic years. Accordingly, it has been possible to determine success in the immediate impact of generating interest and the potential future sustainability of lectures.

Measuring Prevention

However, in the longer term determining the success of preventing scientists (regardless of a terrorist or state affiliation) from contributing to the development of biological weapons is much more difficult. Also, although there may be measures that can be used to assess the long-term impact of biosecurity education, greater consideration is required into how to develop metrics that are sufficiently robust to demonstrate success or failure, yet simple enough to be useful to the policy community. Indeed, it is not enough to simply count the number of students' taught. The National Academies of Science have suggested that experience with other Cooperative Threat Reduction (CTR) activities in the US indicated 'hard "scorecard" metrics, often very quantitative in nature, are not always going to be adequate measures of a programme's success'. [3] This is certain to be the case in relation to biosecurity and dual-use education, where demonstrating the effect in the long term to some extent requires proving a negative without confusing causations.

3 National Academies of Science 2009, *Global security engagement: A new model for Cooperative Threat Reduction*, available: http://www.nap.edu/catalog/12583.html

Lessons Learned

Based on the authors' experience with biosecurity and dual-use education, a number of key practical lessons have emerged. These are not necessarily applicable to other regions and future activities are likely to need tailoring to the specific location. Nonetheless, they may prove useful in facilitating future activities.

One of the principle lessons learned over the course of building the network and conducting implementation tests is the need to carefully frame both engagement and presentation of material. The perception of biosecurity education as irrelevant or less relevant amongst many (not all) in the scientific community, necessitates that the rationale for including it needs to be clearly and objectively articulated. This applies both at university level, where it is essential to ensure science and scientist are not presented as part of the problem but part of the solution, but also to any activity related to other intervention points that could be conducted. Indeed, the need to frame the issue carefully is likely to be even more acute in relation to national academies and authors.

In practical terms, discussion with lecturers and course coordinators in the European context indicates there is only limited space within the degree course syllabi. Thus, space for extensive historical material and detailed information on conventions and regulations is likely to be difficult to integrate within the often-small window of opportunity that exists to engage life-science students. As one survey respondent suggested, 'all knowledge is useful. It is a matter of priorities and of limited number of credits/programme [space available]'. Accordingly, if the objective is to reach a broader audience using the bottom-up approach outlined above, compromise will often be necessary.

Objectivity is also very important to students of disciplines such as biology that are founded on positivist logic. Thus it is often advantageous to avoid subjective approaches or, more significantly, 'preaching' to life-science students. This is particularly acute in relation to more ethically orientated issues such as the morality of using disease in warfare and our experience concurs with the sentiments of Rappert expressed in Chapter 1 that 'resistance can be intense when some try to tell others what they should think'. Finally, there is no one-size-fits-all approach to achieving biosecurity education through a bottom-up approach. A degree of flexibility in the presentation of materials and content is also likely to be important in the future and the model used in the European context is unlikely to be directly applicable to other contexts.

Addressing Other Intervention Points

Although much of the current attention to education is focused on university-level education, this is not the only possible implementation point through which biosecurity education can be progressed and it would be remiss not to address the range of others. In their chapter, Whitby and Dando discuss the role of funders of scientific research in encouraging consideration of the so called dual-use issue. If education is considered an important element of what has been termed the 'web of prevention', [4]there are a number of further intervention points that can and should be addressed which would be mutually reinforcing, examples of which are evident in Figure 14. Indeed, based on a case study produced by Revill, it suggests that if calls for biosecurity education are converted into effective action there are a number of additional intervention points which are particularly important to engage. These include academies of science, authors of scientific textbooks, the biotechnology industry, and possibly schools. [5] Each one of these points is addressed in turn in the following sections.

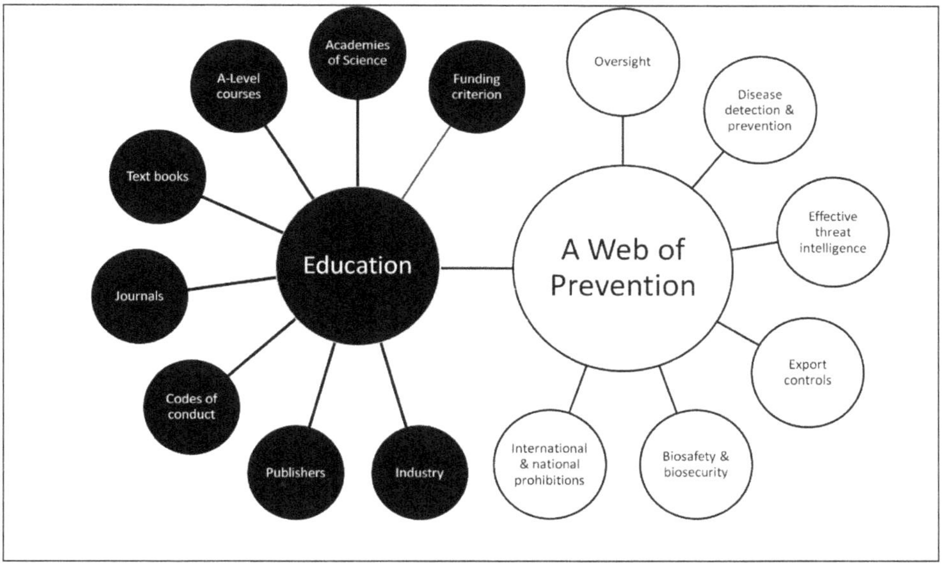

Figure 14: The role of education, and sub activities related to education, within the web of prevention

4 Rappert, B. and McLeish, C. (eds), *A Web of prevention: Biological weapons, life sciences and the future governance of research*, London: Earthscan: 51–65, available: http://people.exeter.ac.uk/br201/Research/Publications/Chapter%203.pdf [viewed 1 November 2009].

5 Revill, J. 2009, '*Biosecurity and bioethics education: A case study of the UK context*', Research Report for the Wellcome Trust Project on 'Building a Sustainable Capacity in Dual-use Bioethics', December 2009, available: http://www.brad.ac.uk/acad/sbtwc/dube/publications/BioseBioethicsUK.pdf.

Academies of Science

Academies of science and 'champions' [6] within the scientific community certainly represent key means of promulgating biosecurity education internally within the community of life scientists, a process that is likely to be considerably more influential than attempts by individuals exogenous to this community. Already there are examples of how individual champions can generate awareness around certain issues, as former UN Secretary General Kofi Annan stated of Joseph Rotblat, 'Mr Rotblat went from working on the nuclear bomb to founding the Pugwash conference, and continued for the rest of his days to champion the principle of scientists taking responsibility for their inventions'. [7] However, in the current climate it seems the life sciences lack such champions and although some national academies of science have demonstrated interest in the area there is insufficient attention placed on this issue to trigger significant activity.

Scientific Literature and Authors

It is also apparent, certainly in the UK, that much (not all) of life-science literature tends to omit discussions on biosecurity. This is understandable given the relatively new nature of these terms in the framework of international security discussions, and the perceived irrelevance of the topic to science and scientist. Similarly, although there is often material on biological weapons, this is within the context of agent characteristics and from a scientific perspective that is devoid of material on the illegality of the use of biology as a weapon. This is also logical given the target audience and the intended purpose of life-science textbooks. Nonetheless, a modest extension could serve to engender further debate on prohibitions and regulations that could raise awareness. There are grounds for optimism in this regard and discussion with textbook authors revealed that some might be willing to include a reference to these issues in future volumes. Certainly, Professor Robert Bauman, author of Microbiology with Disease Taxonomy has stated in discussion with Revill that he plans on 'expanding the content of the section on biological weapons to indicate the illegality of these weapons', believing it 'wise to enhance greater understanding of this important issue'.

---

6 Certainly the idea of 'champions' was voiced at the National Academies of Science [US] meeting on 'Education on Dual Use Issues in the Life Sciences', held on 16–18 November 2009, Warsaw, Poland.

7 Annan, K. 2005, 'Annan regrets death of Nobel Peace Laureate and disarmament advocate Joseph Rotblat', Pugwash Collection of Tributes to Joseph Rotblat, available: http://www.pugwash.org/publication/obits/obit-rotblat-tributes.htm.

The Biotechnology Industry

The role of industry in education could also prove a valuable intervention point for biosecurity education. Like academies of science, the biotechnology industry is likely to have greater sway upon science students than are social scientists or policymakers who, in some cases, are already perceived as overly burdening and constraining scientific research. Moreover, such a process could present benefits and opportunities for corporations to demonstrate a commitment to social and ethical responsibility.

Conclusions: One Approach to the Implementation of Education

In his introduction to this volume, Rappert posited a number of questions that need to be considered in the process of moving biosecurity education from an aspiration to a concrete activity. One such question was: How can audiences of practising scientists or other practitioners be reached and how can their attention and active engagement be secured? Although it would be inappropriate to claim ownership of any definitive response, it is apparent that the process undertaken through the LNCV–BDRC study provides a useful overview of the extent of education, but also an understanding of the attitudes of life-science educators and a collection of contacts necessary to construct a collaborative network. This can have practical value as a vehicle to conduct implementation tests and convey ownership of the process of biosecurity education to life scientists. Although the efficacy of this approach in the context of other cultures remains contestable and further research and adaptation may be required to tailor education to the local context, in Europe and other Western academic environments this network process is one constructive bottom-up model through which an audience can be reached and their attention and active engagement secured. Indeed, the network has proved useful in conducting implementation tests in academic departments around Europe and learning to tailor the material and delivery of the EMR.

It has further demonstrated there is no one-size-fits-all approach to education even within the framework of European universities; and although a core module resource is a useful basis for implementation, the context needs to be taken into account. Indeed, based on lessons learned over the course of this project, it is essential to recognise the various priorities of different communities of practice and ensure that material is delivered objectively in a manner that avoids preaching and balances the potentially harmful effects of life-science research with the vast array of positive benefits generated by scientific research.

This is likely to be a greater challenge in cases where the model was applied more broadly within other contexts, particularly those where scientific research assumes a much higher priority than security of scientific research.

Finally, the implementation of biosecurity at university level is only one element of a number of measures that could be taken to build a 'culture of responsibility'. Other intervention points, such as academies of science and 'champions', the biotechnology industry, and scientific authors, all have a significant role to play advancing biosecurity education. As many of the questions relating to the 'who, what and how of education' will ultimately be answered by the scientific community, significant input and engagement with these other intervention points will be essential in the long term.

Chapter 10: Biosecurity Awareness-raising and Education for Life Scientists: What Should be Done Now?

SIMON WHITBY AND MALCOLM DANDO

Introduction: Awareness

The 1975 Biological and Toxin Weapons Convention (BTWC) added to the ban on the use of biological weapons embodied in the 1925 Geneva Protocol by what was termed the 'General Purpose Criterion' of Article 1. This stated that: [1]

> Each State Party to this Convention undertakes never in any circumstances to develop, produce, stockpile or otherwise acquire or retain:
>
> 1. Microbial or other biological agents, or toxins whatever their origin or method of production, of types and in quantities that have no justification for prophylactic, protective or other peaceful purposes...

Thus, peaceful uses of the modern life sciences are fully protected, but there is an all-encompassing prohibition of non-peaceful development, production, stockpiling, acquisition or retention of microbial or other biological agents or toxins (and toxins here are understood to cover all mid-spectrum agents such as bioregulators).

As early as the Second Five-Yearly Review Conference of the BTWC in 1986, the 'States Parties' [2] recognised the importance of the awareness and education of life scientists in regard to the Convention. In the Final Declaration of

1 For the text of the Convention, see: www.opbw.org.

2 The term 'States Parties' refers to the membership of the Convention. The Biological Weapons Convention currently has 163 States Parties and 13 signatories. There are 19 states which have neither signed nor ratified the Convention.

the Conference, States Parties [3] noted, in relation to Article IV on national implementation measures, that: [4] 'The Conference notes the importance of ...inclusion in textbooks and in medical, scientific and military educational programmes of information dealing with the prohibition of microbial or other biological agents or toxins and the provisions of the Geneva Protocol.'

Similar statements were subsequently agreed at following Review Conferences. However, during the 1990s, when concerns about Biological Warfare (BW) returned, the attention of diplomats was centred on the problem of how confidence in compliance with the Convention might be improved and there was very little involvement of the civil life-science community. [5] After the failure of these efforts, when states decided to discuss and promote common understandings on more tractable issues, Australia reported in a 2005 Intercessional Meeting on Codes of Conduct that: [6]

> 1. Amongst the Australian scientific community, there is a low level of awareness of the risk of the misuse of the biological sciences to assist in the development of biological or chemical weapons. Many scientists working in 'dual-use' areas simply do not consider the possibility that their work could inadvertently assist in a biological or chemical weapons programme...

At the same meeting we reported work carried out with the editor of this volume, Brian Rappert, in which we had interactive seminar discussions with life scientists at 15 UK universities. Analysis of the tape recordings of these seminars led us to conclude that: 'There is little evidence from our seminars that participants: a. regarded bioterrorism or bioweapons as a substantial threat; b. considered that developments in life-sciences research contributed to biothreats; c. were aware of the current debates and concerns about dual-use research; or d. were familiar with the BTWC.' [7]

In the next year we reported to the Sixth Review Conference on further seminars in several other countries. In regard to the UK seminars we concluded that: 'The results from the remainder of the seminars were consistent with all

---

3 States Parties refers to States that have both signed and ratified the Biological and Toxin Weapons Convention. (BTWC). The BTWC currently has 163 States Parties and 13 signatories. There are 19 states which have neither signed nor ratified the Convention.

4 Again, for the text of the Final Declaration, see: www.opbw.org.

5 Dando, M. R. 1994, *Biological warfare in the 21st century: Biotechnology and the proliferation of biological weapons*, London: Brasseys; Dando, M. R. 2002, *Preventing biological warfare: The failure of American leadership*, Basingstoke: Palgrave.

6 Australia 2005, *Raising Awareness: Approaches and Opportunities for Outreach*, BWC/MSP/2005? MX/WP.29, available: www.opbw.org.

7 Dando, M. R. and Rappert, B. 2005, *Codes of Conduct for the Life Sciences: Some Insights from UK Academia*, Briefing Paper no. 16 (2nd series), University of Bradford, May, available: www.brad.ac.uk/acad/sbtwc.

of these points [enumerated above]. A particular surprise was that so few of the participants (less that 10 per cent in most groups) had heard of the mousepox experiment that has figured largely in security literature.' [8]

There were, of course, some differences in the interactions that we reported in the seminars in the Netherlands, Finland, the US and South Africa but we stated that '[D]espite such differences between the seminars held in the different countries, the degree of similarity between the responses in the seminars was much more pronounced'. Our subsequent experience of carrying out seminars in 16 different countries with a few thousand life scientists in over 110 different departments has consolidated these findings. [9] Indeed, we used the seminars more as an awareness-raising mechanism rather than as a means of investigative research into the attitudes of life scientists.

The Education Gap

These findings demand some serious explanation. Many physicists are clearly aware of the dangers of the misuse of their science and have played important roles, for example, in the Pugwash movement. In the 1980s and 1990s, chemists were also influential in helping to bring negotiations of the Chemical Weapons Convention (CWC) to a successful conclusion, and the International Union of Pure and Applied Chemistry (IUPAC) has contributed major reviews of relevant science and technology to the first two Review Conferences of the CWC. Therefore, it is not unreasonable to ask why practising life scientists are so unaware of the BTWC and the problem of dual use despite increasing attention being given to these issues, for example, by national science academies.

One possible explanation is that life scientists are uninformed of biosecurity issues because they do not feature in their university education. In order to investigate this possibility, in cooperation with the Italian Landau Network Centro-Volta, we carried out an internet survey of a sample of courses in the EU. As detailed further in the chapter by Mancini and Revill, the results were quite startling: 'This research suggested that only three out of 57 Universities identified currently offered some form of specific biosecurity module and in all cases this was optional for students.' [10] On the other hand, the survey noted:

---

8 Rappert, B., Chevrier, M. I. and Dando, M. R. 2006, *In-depth Implementation of the BTWC: Education and Outreach*, Review Conference Paper no. 18, University of Bradford, available: www.brad.ac.uk/acad/sbtwc.

9 Rappert, B. 2007, *Biotechnology, security and the search for limits: an inquiry into research and methods*. Basingstoke: Palgrave; Rappert, B. 2009, *Experimental secrets: International security, codes, and the future of research*, University Press of America.

10 Mancini, G. and Revill, J. 2008, *Fostering the biosecurity norm: Biosecurity education for the next generation of life scientists*, London Network, Centro Volta and University of Bradford, available: www. Dual-use bioethics.net.

> There is evidence of a considerable number of bioethics modules and nearly half of the degree programmes surveyed evidenced some form of bioethically-focused module. In terms of biosafety modules... roughly one-fifth of life-science degrees in the sample contain a specific dedicated biosafety module although several of these specific modules were optional.

So we found a reasonable number of biosafety modules, a large, and we suspect, increasing number of bioethics modules, and virtually no biosecurity modules.

We attempted to investigate in more detail by looking for any kind of reference to biosecurity issues in the course material. Again the picture was bleak:

> Exactly what constitutes a reference varies; however, based on the quantitative data from the investigation, we found a total of 37 life-science degree courses out of our sample of 142 where there was clear evidence of a reference to biosecurity. Only a minority of the degree courses in the study — a total of 22 out of 142 — made a reference to the BTWC, BW and/or arms control, and a similar number, 29 degree courses, exhibited some reference to the dual-use issue.

When we carried out a similar survey in Japan, and as can be seen from the analysis presented by Minehata and Shinomiya, we found a similar picture. Of 197 life-science degree courses in 62 universities we found only three specific biosecurity modules. [11]

In Japan we took the investigation a stage further by sending out a questionnaire to lecturers asking why biosecurity and dual use was not being taught. Clearly some lecturers did not see these subjects as relevant to their courses, but others certainly did. Where people thought the topics relevant but did not teach them, the reasons cited were a lack of expertise and access to necessary resources, and a lack of space on a very crowded timetable in the modern life sciences.

Correcting the Deficiency

Correcting this deficiency in education- and awareness-levels of life scientists will be a massive task that will require action by a range of constituencies involved in life-science education including, *inter alia*, governments, bodies

11 Minehata, M. and Shinomiya, N. 2009, *Dual-use education in life science degree courses at universities in Japan*, National Defence Medical College of Japan and Bradford University, available: www.dual-use bioethics.net.

responsible for the administration of standards in higher education, funders of life-science education, civil society groups and non-governmental organisations involved in the production of educational material, and teachers and trainers. [12]

As is evidenced by the convergence of ethics and medicine in the area of biomedical ethics, [13] the consideration of moral dilemmas is not new in life-science research. However, current concern about dual use [14]— where science findings can be used for malign as well as benign purposes — arises from a new range of security threats. These include the changing nature of warfare, the possibility of new forms of mass-casualty terrorism, a discernable commitment by States Parties to the BTWC to address these threats through seeking to improve awareness and education amongst life scientists, reviews of scientific oversight regarding dual-use research performed by national scientific academies [15] (particularly in the US), and a genomic and biotechnology revolution in life science with the rapid and worldwide spread of advanced science and technology. Thus concerns about dual use are being discussed in the context of a distinctly new phenomenon — namely, a convergence between security concerns and the practice of life scientists in what might be termed a novel biosecurity problem.

The term 'biosecurity' has been used in different ways in different contexts. We should, therefore, be very clear about usage of the term here. In our view the threat spectrum ranges from natural disease through to inadvertently caused disease. We deal with natural disease by public-health measures and inadvertently caused disease is restricted by 'biosafety' — good laboratory practice. The concept of laboratory biosecurity has also arisen to ensure that dangerous materials are kept secure from those with malign intent. We see laboratory biosecurity as part of biosecurity, but for us the term has a much wider meaning related to the concept of a web of preventative policies centred on the prohibition of the misuse of the life sciences embodied in the General Purpose Criterion of the BTWC. Thus biosecurity is the objective of the whole range of policies, such as export controls, biodefence and national implementation of the Convention, that minimise the possibility that the life sciences will be misused for hostile purposes. Within that range of policies there is, in our opinion, a role for practising life scientists in being aware that the materials, technologies and knowledge they produce may be misused and for contributing their expertise to the development and maintenance of preventative policies.

---

12 Dando, M. R. 2009, Dual-use Education for Life Scientists, *Disarmament Forum*, vol. 1, pp. 41–4.

13 Jonsen, A. R. 1998, *The birth of bioethics*, 1st edition, USA: Oxford University Press.

14 Atlas, R. and Dando, M. 2006, 'The dual-use dilemma for the life sciences: Perspectives, conundrums, and global solutions', *Biosecurity and Bioterrorism: Biodefense Strategy, Practice, and Science*, vol. 4, pp. 276–86.

15 The US National Academy of Sciences' (Fink Committee) classification of seven classes of experiment sought to illustrate the types of endeavour that would require careful review by informed experts.

In order to begin building capacity in biosecurity education, a new range of creative and innovative interventions are required. As set out in the following argument, it is seen that biosecurity education can be easily accommodated by current standards in higher education in the UK. In the US, recommendations for the adoption by federally funded institutions of biosecurity education have already been set out.

Scope for Biosecurity Education in the UK and US

Whilst ethical consideration of the implications of dual-use science and technology is conspicuously absent from the vast majority of curricula in UK higher education, in the US, and indeed worldwide, it is apparent that a codified response through the development of new guidelines and policies that reflect biosecurity concerns will not necessarily be required within UK higher education. In order to satisfy its statutory obligation to ensure that publicly funded teaching provision is of a high standard, the UK's Higher Education Funding Council (HEFCE) [16] contracts the Quality Assurance Agency (QAA) [17] to 'devise and implement quality- assurance methods' and is responsible for the conduct of audit and review of teaching quality in both higher and further education. Although not a national curriculum that sets standards in UK higher and further education, the Subject Benchmark Statements produced by the Quality Assurance Agency [18] require the inclusion of an ethical dimension in all undergraduate bioscience programmes. These are largely aspirational; however, they 'set out expectations about standards of degrees in a range of subject areas. They describe what gives a discipline its coherence and identity, and define what can be expected of a graduate in terms of the abilities and skills needed to develop understanding or competence in the subject.' [19]

The 2002 QAA bioscience benchmark statements made a number of references to 'ethical' aspects of this subject-area, including the following requirements:

> Students should expect to be confronted by some of the scientific, moral and ethical questions raised by their study discipline, to consider viewpoints other than their own, and to engage in critical assessment and intellectual argument. [20]

---

16 http://www.hefce.ac.uk/Learning/qual/qaa.asp.

17 http://www.qaa.ac.uk/academicinfrastructure/benchmark/default.asp.

18 UK Quality Assurance Agency (QAA) 2002 & 2007, *Honours Degree Subject Benchmark Statements*, Bioscience, available: http://www.qaa.ac.uk/academicinfrastructure/benchmark/statements/Biosciences07.pdf.

19 Ibid.

20 Ibid.

> Recognising the moral and ethical issues of investigations and appreciating the need for ethical standards and professional codes of conduct. [21]
>
> All students should: Have some understanding of ethical issues and the impact on society of advances in the biosciences. [22]
>
> Good students should: Be able to construct reasoned arguments to support their position on the ethical and social impact of advances in the biosciences. [23]

Honours Degree Subject Benchmark Statements were re-stated by QAA again in 2007, where the Subject Benchmark Statement for Biosciences again reiterated the importance of the inclusion of an ethical dimension in undergraduate programmes. Whilst ethics teaching forms an important component in many bioscience courses and courses address a range of ethically related issues, ethics in bioscience or bioethics education could easily be extended to accommodate and incorporate the ethical concerns of biosecurity education. Institutional audits of teaching-quality assessment by the QAA strengthen incentives to extend teaching in ethics into the area of biosecurity.

Further to this, a report [24] published in December 2008 by the US National Advisory Board for Biosecurity (NSABB) sets out a Strategic Plan for Outreach and Education on Dual-Use Research Issues. As specified in a related contribution to this book, this strategy envisages the implementation of a series of recommendations on 'the development of programmes for outreach, education, and training on dual-use research issues for all scientists and laboratory workers at federally funded institutions in the US'.

Funders of Science

Increasingly, recipients of research funding must be willing to comply with requirements set out by funders of science that are intended to ensure bioscience-research activities are in full compliance with guidance on ethics. Indeed, in the case of some funders of bioscience, reference to dual-use research is now explicit.

---

21 Ibid.

22 Ibid.

23 Ibid.

24 National Science Advisory Board for Biosecurity 2008, *Strategic plan for outreach and education on dual use research issues*, December, Washington, DC.

As outlined in its position statement [25] on Research Ethics, the UK Biotechnology and Bioscience Research Council (BBSRC) states that it has:

> a responsibility to ensure that its funds are used ethically and responsibly. Potential applicants should consider whether their work is likely to give rise to societal concerns about the purpose of the research, or includes any social or ethical issues regarding its conduct or potential outcomes (for example, relevance to development of biological weapons; products and processes that might be used in social discrimination), or other aspects of potential public concern.

As stated in its Terms and Conditions [26] for Research Council Grants, according to the Research Councils UK (RCUK), recipients of its funding are responsible for ensuring that:

> ...ethical issues relating to the research project are identified and brought to the attention of the relevant approval or regulatory body. Approval to undertake the research must be granted before any work requiring approval begins. Ethical issues should be interpreted broadly and may encompass, among other things, relevant codes of practice, the involvement of human participants, tissue or data in research, the use of animals, research that may result in damage to the environment and the use of sensitive economic, social or personal data.

The UK Medical Research Council (MRC) [27] sets out detailed guidance on ethics that addresses a broad range of areas including 'clinical research governance', 'global bioethics' and 'good research practice'. Additionally, the MRC Position Statement on Bioterrorism and Biomedical Research recognises the 'dual-use nature of life science and the importance in funding research of due consideration of ethical dilemmas presented by research'.

Adopting a similar approach, the Wellcome Trust specifies the importance of appropriate processes existing at institutional, national and international levels for the review and oversight of dual-use research. In its Position Statement on Bioterrorism and Biomedical Research, the Wellcome Trust [28] cites the US National Academy of Sciences' (Fink Committee) classification of seven classes of experiment to illustrate the types of endeavour that would require careful review by informed experts. The experiments this committee specified are those that would:

- demonstrate how to render a vaccine ineffective

---

25 http://www.bbsrc.ac.uk/funding/apply/grants_guide.pdf.

26 http://www.rcuk.ac.uk/cmsweb/downloads/rcuk/documents/tcfec.pdf.

27 http://www.mrc.ac.uk/Ourresearch/Ethicsresearchguidance/index.htm.

28 http://www.wellcome.ac.uk/About-us/Policy/Policy-and-position-statements/WTD002767.htm.

- confer resistance to therapeutically useful antibiotics or antiviral agents
- enhance the virulence of a pathogen, or render a non-pathogen virulent
- increase transmissibility of a pathogen
- alter the host range of a pathogen
- enable the evasion of diagnostic and detection modalities
- enable the weaponisation of a biological agent or toxin.

Improved grant-application procedures, more stringent reporting requirements and management, and oversight of the grants by funders will help ensure that grantees live up to stipulated ethical obligations and efficient and effective implementation of such requirements will facilitate the development of best practice in the financing of life-science research. Given the low levels of awareness amongst life scientists of biosecurity issues it is not surprising that progress has been slow in the implementation of measures such as effective training programmes to address the issue. However, it can be expected that these measures will be implemented in coming years: the question is, how quickly and how well?

Civil Society Groups and Non-Governmental Organisations

In the meantime, there does seem a potential role for civil society in providing models of what might be done to close the gap most effectively in the shortest timeframe. This is what we have been attempting to do over the last few years in developing a Dual-use Biosecurity Education Module Resource (EMR). The NSABB report 'Strategic Plan for Outreach and Education on Dual Use Research Issues' [29] mentioned previously considers what needs to be done in some detail. In its view, developing a strategic plan requires: 'First and foremost, the target audience must be identified and assessed as to their level of understanding of the issues since this will guide educational strategies...[Then]...messages should be tailored to specific target-audiences. Key points must be identified and specifically crafted.' And because there are so many different possible methods of communication, 'it is important to select those methods that will most effectively reach the intended audiences'.

Therefore, when we applied a similar method of analysis to our work, it was clear that our intended target audience — university-level lecturers and students — did not have a high level of awareness of biosecurity and dual-use issues. Furthermore, given the prevalence of the use of the internet in universities it

29 National Science Advisory Board for Biosecurity 2008, op. cit.

was clear that providing information on the web was by far the most efficient and effective way forward. However, given the pressure on the timetable we thought it unwise to design a one-size-fits-all educational module and decided to design an EMR that could be used by different lecturers to fit relevant parts into their own courses.

Our thinking was also much influenced by the developing consensus about education of life scientists that developed at the 2008 BTWC Intercessional Meetings. The final report of these meetings states: [30]

> 26. States Parties recognised the importance of ensuring that those working in the biological sciences are aware of their obligations under the Convention and relevant national legislation and guidelines, have a clear understanding of the content, purpose and foreseeable...security consequences of their activities, and are encouraged to take an active role in addressing the threats posed by potential misuse of biological agents and toxins as weapons, including bioterrorism.

This paragraph of the report then continues, significantly: 'States Parties noted that formal requirements for seminars, modules or courses, including possible mandatory components, in relevant scientific and engineering training programmes and continuing professional education could assist in raising awareness and in implementing the Convention.'

In the paragraph that followed, States Parties set out what they agreed would be of value in such programmes:

> (i) Explaining the risks associated with the potential misuse of the biological sciences and biotechnology
>
> (ii) Covering the moral and ethical obligations incumbent on those using the biological sciences
>
> (iii) Providing guidance on the types of activities which could be contrary to the aims of the Convention and relevant national laws and regulations and international law
>
> (iv) Being supported by accessible teaching materials, train-the-trainer programmes, seminars, workshops, publications, and audio-visual materials

30 Meeting of the States Parties to the Convention on the Prohibition of the Development, Protection and stockpiling of Bacteriological [Biological] and Toxins Weapons and on their Destruction (2008) Report of the meeting of States Parties, BWC/MSP/2008/5, United Nations, Geneva, 10 December.

(v) Addressing leading scientists and those with responsibility for oversight of research or for evaluation of projects or publications at a senior level, as well as future generations of scientists, with the aim of building a culture of responsibility

(vi) Being integrated into existing efforts at the international, regional and national levels.

Our idea for the EMR was to capture as many of these ideas as possible based on the concept of having a web of integrated preventative policies that together would persuade everyone thinking of breaking the prohibition that the costs would far outweigh the benefits. [31] However, in work with colleagues at Japan's National Defence Medical College on designing and testing the EMR (under British Council Funding) it became clear that we needed to start the lecture series with material that could be readily grasped by life scientists.

Thus the EMR consists of 21 lectures, each with 20 PowerPoint slides and notes for the lecturer, and direct links to the references used via the web. Each lecture also has some suggested essay questions and the EMR has an introduction to all the material for lecturers and a small number of Briefing Papers cover material that would be less familiar to life scientists. Several lectures are also duplicated, with material in the second set being more scientifically orientated.

Therefore, our EMR is designed in five parts as follows:

Outline of the EMR

A. Introduction and Overview

Lecture 1

B. The Threat of Biological Warfare and Biological Terrorism and the International Prohibition Regime

Lectures 2–10

C. The Dual-Use Dilemma and the Responsibilities of Scientists

Lectures 11–18

D. National Implementation of the BTWC

Lectures 19–20

E. Building an Effective Web of Prevention

---

31 Rappert, B. and McLeish, C. 2007, *A Web of prevention: Biological weapons, life sciences and the governance of research*, London: Earthscan.

Lecture 21

Thus the first lecture gives a brief overview of the whole of the module resource in order to orientate the user.

The second section takes up the story of the misuse of modern biology after the discovery of the causes of infectious diseases in the late nineteenth century by scientists such as Pasteur and Koch. This history is largely unknown amongst life scientists and forms a basis for consideration of the possible misuse of future advances. In this section we have also introduced modern accounts of the traditional agents such as anthrax, smallpox and botulinum toxins to better engage scientists' interest.

The lectures in Section B are set out as follows:

Section B of the EMR

2. BW from Antiquity to World War I

3. BW from WWI to WWII

4. BW during the Cold War

5. The impact of BW Agents

6. Assimilation of BW in State Programmes

7. International Legal Agreements

8. Strengthening the BTWC 1980–2008

9. The 2003–2005 Intercessional Process

10. The 2007–2010 Intercessional Process

Section B ends by briefly reviewing how the international community has attempted to deal with the threat of the proliferation of biological weapons through the 1925 Geneva Protocol, the 1975 BTWC and the 1997 CWC (given that there is an overlap between the BTWC and CWC in the area of mid-spectrum agents such as toxins and bioregulators).

It can be seen that these lectures begin with a consideration of the history of biological warfare and end with the BTWC recent annual meetings in which scientists have become increasingly involved — at least at the level of national academies and industrial leaders. This then sets the basis for the third section of our module.

The lectures in Section C are set out as follows:

Section C of the EMR

11. Bioethics methodology
12. Obligations derived from the BTWC
13. The growth of dual-use bioethics
14. Dual-use: The Fink Report
15. Dual-Use: Examples
16. The Lemon-Relman Report
17. Weapons targeted at the nervous system
18. Regulation of life sciences

Although present evidence strongly suggests there is little biosecurity or dual-use content in university life-science modules dealing with bioethics, it is our belief this is probably the best place to focus on these issues. Life scientists are becoming familiar with the ethical problems that new research brings up, and the teaching of bioethics is growing in universities. Our view is that biosecurity and dual-use issues are best presented to life scientists in the context of the moral and ethical implications of research (see item (ii) in the 2008 report of the BTWC meeting on education above). [32] Therefore, this section of the module starts with a review of standard bioethical analyses that students are likely to have encountered before, introducing the growing literature on dual-use bioethics. The section then leads to a consideration of the key US National Academics Reports (Fink and Lemon-Relman) that began the closer examination of the dual-use problem from within the scientific community. Some lectures examine classic dual-use experiments such as the mousepox experiment in lecture 15 and the contention by Lemon-Relman that the dual-use problem is far wider than just research in microbiology, as illustrated in lecture 17 regarding concerns over the misuse of advances in neuroscience. The section ends with a lecture that reviews the various papers that have recently discussed regulation of the security implications of the life sciences.

The final two sections of the EMR continue this theme of national and international regulation and are set out as follows:

Section D of the EMR

19. International regulation of biotechnology

32 Revill, J. 2009, *Biosecurity and bioethics education: A case study of the UK context*, Wellcome trust Dual-Use Bioethics Group, University of Bradford, available: www.dual-usebioetics.net.

20. National implementing legislation

Section E of the EMR

21. The web of prevention

Looking back at the list of specific suggestions agreed by State Parties to the BTWC in 2008, we would argue that we have covered most. Lectures cover the risks of misuse, the moral and ethical obligations of life scientists, give guidance on the types of activities which could be contrary to the aims of the Convention, and provide accessible teaching materials. Therefore, what else needs to be done?

Increasing Efficiency through Networks

One way to build on the work described here is to carry out more surveys of education provision in the university sector in different countries. These surveys, particularly if carefully followed up by questionnaire, telephone and email, inevitably provide a list of life-science lecturers who are interested in bringing issues of biosecurity and the dual-use dilemma into their courses. By assisting the development of country and regional networks on the basis of these contacts it should be possible to generate a much faster development and uptake of material suitable for different countries and regions. Such an approach would also fit with the States Parties agreement on the value of education efforts being integrated into existing international, regional and national activities.

As shown in a recent report from the US National Academies on 'Ethics Education and Scientific and Engineering Research: What's Been Learned? What Should Be Done?' dual-use bioethics developments will fit within a broader effort to develop ethics education.[33]

What is also clear is that these wider developments, whilst showing some advances in understanding how to proceed best in engaging students, have not yet found an adequate means of evaluating the impact of teaching on later ethical behaviour. The report points out that:

> Attempts to evaluate and improve ethics education for scientific and engineering research and practice are just beginning. However, they do show that even though immediate results of some programmes are

33 National Academies 2009, *Ethics Education and Scientific and Engineering Research: What's Been Learned? What Should Be Done*? Summary of a workshop, Washington, DC: National Academies Press.

positive, circumstances and pressures can overwhelm graduate students, postdoctoral fellows, and junior-faculty and researchers and undermine results.

In the longer term, attention to evaluation of the impact of dual-use bioethics education will be central to supporting the prohibition embodied in the BTWC.

More immediately, a further chance of improving efficiency will arise at the 2011 Seventh Review Conference of the BTWC because just after the specific suggestions on education (and codes of conduct) discussed above, paragraph 31 suggested that: 'State Parties are encouraged to inform the Seventh Review Conference of, *inter alia*, any actions, measures or other steps they may have taken on the basis of the discussions...in order to facilitate...decisions on further action.'

Therefore, if networks of life scientists concerned with implementing dual-use bioethics education can be established in different countries and regions, and if they carefully evaluate their efforts, the results could be applied rapidly elsewhere to help quickly close the education gap.

What Should Be Done Now?

Even if all of what has been discussed in this chapter were achieved it would still leave a great deal needing to be done. One specific point in the 2008 agreement amongst States Parties to the BTWC seems particularly important to us: train-the-trainer programmes, being an important capacity-building initiative in developing a worldwide culture of responsibility amongst life scientists. In regard to this objective, we believe rapid progress can be made through the use of modern technology.

Train-the-Trainer

In order to facilitate efficient and effective engagement across a range of life-science constituencies worldwide we developed an expert-level online distance-learning train-the-trainer programme in dual-use bioethics (biosecurity) education. The original iteration of this module consisted of two key elements: 1. the EMR described above, together with 20 expert-level scenarios that introduce users to examples of the complex bioethical dilemmas that have confronted life-science research; and 2. a range of innovative electronic online distance-learning technologies that facilitate outreach on a worldwide basis.

The aim of the module was to introduce educators to the concepts in bioethics and biosecurity education by developing awareness and understanding of a range of dual-use ethical issues arising from the impact of science and technology on society. The module provided the opportunity to develop knowledge of approaches that give a defence for ethical decisions or recommendations regarding dual-use technologies. Educators were guided through the lecture series by a trainer. Participants were introduced to scenarios where the results of well-intentioned scientific research can be used for both good and harmful purposes which have given, or may in future give, rise to what is now widely known as the 'dual-use dilemma', providing the opportunity to analyse in depth the ethical dilemmas these scenarios raise. Central to this is the importance and role of ethics in informing the debate. The programme was intended to have an applied, practical dimension in that its aim was to enable and facilitate more bioethical research into dual-use issues, and help develop policies and practices that might prevent the misuse of knowledge generated through biomedical research.

The methodological approach relating to the delivery and implementation of this module was developed with UK academic standards in mind, so that the module would furnish participants with knowledge and understanding to review and appraise ethical theories and methods relevant to dual-use bioethics and recognise and discuss ways in which the application of ethics methodologies resolves or leaves unresolved questions relating to dual-use issues. In relation to subject-specific skills, the module would facilitate educators' organisation and synthesis of ideas and questions relevant to assessing ethical dilemmas in specific dual-use issues affecting humans, animals and plants generally, and across a select range of life-science sub-disciplines of relevance including human biology, zoonotic diseases, phytopthology, biotechnology, DNA synthesis, drug control, genomics, genetic engineering and genetic modification, immunology, nanotechnology, neuroscience, scientific freedom, synthetic biology, whistleblowing, and processes relating to ethical review. As to personal transferable skills, educators would be able to evaluate and integrate data from a variety of sources and express ideas clearly, both verbally and in writing; and communicate effectively in an online environment using a range of media.

Technologies

The module was designed to facilitate participation in lectures, seminars, and discussion groups that would all take place online. A novel approach to online distance learning was adopted. This utilised online distance-learning technologies that facilitate the delivery, viewing of, and participation in lectures by real-time video transmission. With this approach participants can see a live video transmission of the teacher, and the teacher can guide the participants through respective online sessions with the support of a range of online teaching technologies and visual aids. Together with the lecture, these can be broadcast simultaneously, including PowerPoint presentations, word-processing files, graphical images, as well as audio and video. Participants with video cameras can be invited by the teacher to join live online 'face-to-face' discussion and the latter can be viewed online by all members. Those with the capability to transmit audio can raise a (virtual) hand, be invited by the teacher to join live online discussion, and can communicate this way with all of those taking part, regardless of geographical location. Participants without video and audio capability can follow the class online and communicate with the teacher by typing questions via their keyboard. The classes can be recorded and viewed online subsequently.

Working in a fully supported online-learning community, members are able to communicate and interact with peers, developing their practice through sustained reflection and involvement in a range of activities and scenarios. Participants are encouraged to bring their own ideas and experiences to the course, sharing these with peers to contextualise their knowledge and understanding in ways that will help them, as life-science professionals, to meet the ethical challenges thrown up by dual use. As well as participation in a vibrant academic (social-network) web-group where interaction on coursework-related topics between tutors, moderators and students takes places, members undertake independent reading and research. Participants benefit from a supportive and interactive online web-based learning community and work both independently to produce a coursework assignment, as well as in online groups to produce a significant group-work course assignment.

Conclusion: Recommendations for the BTWC

As has been made clear, a major effort will be required to raise awareness levels amongst life scientists and develop a culture of responsibility around the dual-use implications of research. Whilst concerns raised by high-level reviews of scientific oversight of dual-use research are becoming assimilated into the terms and conditions associated with the funding of life-science research, this is a long-term initiative and will necessarily involve a broad range of constituencies. As set out by Mathews and Webb, [34] two practical suggestions would assist in sustaining interest in this area. The first is that States Parties could report to the Seventh Review Conference of the BTWC in 2011 on progress on implementation and capacity building in dual-use/biosecurity education. The second is that

34 Mathews, R. J. and Webb, J. M. 2009, 'Awareness-raising, Education and Codes of Conduct within the Framework of the BWC', Chapter 9, *BWPP Biological Weapons Reader*, McLaughlin, K. and Nixdorff, K. (eds).

the confidence-building mechanism (CBM) could be extended by appending progress reports on implementation and capacity building in education for life scientists to annual CBM reports.

Chapter 11: Teaching Ethics to Science Students: Challenges and a Strategy

JANE JOHNSON

To be an effective scientist in the twenty-first century requires not only a specialised scientific knowledge but an appreciation of the ethical dimension of science. Scientists need to be able to recognise ethical dilemmas and formulate coherent responses to them. But scientists are not philosophers or ethicists, and their ethics education, therefore, needs to be different from that frequently offered as part of mainstream ethics courses, particularly those on moral theory. This chapter will argue that dual-use dilemmas and role-play involving real scientific case studies are an ideal vehicle for effectively engaging future scientists in ethics education, and helping furnish the necessary skills for their professional development.

The Role of Ethics Education for Scientists

Two questions which ought to precede any properly informed discussion of how to teach ethics to scientists are 'Why should we teach this group ethics?' and 'What do we hope to achieve from their ethical education?' Ethics teachers who are novices in the area might well be driven to ask these questions in despair as they confront resistance to their efforts on the part of both students and their colleagues in the science faculty. Nonetheless, how we respond to these questions is a serious matter and crucial to determining the shape of ethics courses.

A recent workshop on ethics education in science and engineering began by asking participants why they thought ethics education was important. Respondents talked about famous cases of research misconduct (presumably hoping they could be prevented in the future by ethics education) and how public trust in the integrity of science and research may be undermined by problematic practices. It was also noted that some students only appreciated the value of their ethics education in retrospect, after practising their discipline and being forced to confront real-life ethical issues. Interestingly, there was also a suggestion that talented students with high ideals might be lost if ethics education were ignored.[1] Whilst all these factors can play a role in motivating ethics education for scientists, the central problem which surely underpins them all is that ethical issues constantly arise in science, and scientists need to learn how to deal with them. As researchers investigating ethics education in the life sciences have noted, the 'more influential science becomes, the more ethical issues become associated with scientific practice directly, and scientists are increasingly required to participate in the value questions born from new knowledge and new technologies'.[2]

There are a number of ways in which the practice of science generates ethical issues. Regarding the methods adopted in research (for instance, we can ask 'Should we run placebo-controlled drug trials, or use animals in experimentation?') these include how knowledge is applied (for example, how do we respond to knowledge of aerosolisation being used to make more effective bioweapons?), as well as the very questions which drive scientific research in the first place (for example, should we do research into human reproductive cloning, or weapons of mass destruction?). In fact, the ethically charged nature of science is well exposed by the dual-use dilemma, since dual-use scenarios demonstrate that even the well-intentioned pursuit of scientific research can generate difficulties. Although a scientist may be pursuing admirable goals such as understanding how a particular disease spreads with a view to containing future outbreaks, this does not preclude this same research being used for harmful ends such as deploying the disease as a biological weapon.

Since WWII at least there has been a growing awareness that the ethical challenges generated by science need to be addressed. To this end various strategies have been tried, including codes of conduct and ethics such as the World Medical Association's Declaration of Helsinki and the International Ethical Guidelines for Biomedical Research Involving Human Subjects; regulations and laws like those prohibiting research on human reproductive cloning in Australia; and boards or committees in universities, hospitals and other institutions charged with assessing whether proposed scientific research is ethical.

Though motivated by laudable goals, these measures are limited and for a variety of reasons fall short of what is required to address the ethical challenges generated by science. For instance, one issue with respect to some codes of conduct and ethics is that although they may supply aspirations and even

---

1 Hollander, R. (ed.) and Arenberg, R. A. (co-ed.) 2009, *Ethics education and scientific and engineering research: What's been learned? What should be done?*, Washington, DC: The National Academies Press, p. 6.

2 Clarkeburn, H., Downie, J. R. and Matthew, B. 2002, 'Impact of an ethics programme in a life sciences curriculum', *Teaching in Higher Education*, vol. 7(1), pp. 65–79.

appropriate rules for behaviour, individuals may not possess the practical know-how or skills to apply them.[3] Thus, even the most well-conceived, well-written and comprehensive codes may not foster ethical conduct if the individuals to whom they are meant to apply do not understand how to follow them, or have not been involved in the process of developing them.[4] When it comes to laws and regulations, these are frequently backward-looking and thereby may be ill-equipped to cope adequately with new situations being generated by science. In the case of ethics committees, their purview is limited. They are generally constrained to monitoring human and nonhuman animal experimentation rather than research in fields like physics, chemistry and engineering, and are charged with determining whether experiments that have already been conceived and developed comply with an institution's policies. Finally, none of these measures genuinely address the institutional and cultural factors that may impede ethical conduct in science.[5] Therefore, it seems that if we as a society take ethics in science seriously, we do our future scientists a disservice if we do not adequately prepare them for the ethical challenges they will face, since we cannot depend solely on outside parties and existing mechanisms to ensure 'good' science.[6]

In accepting that scientists need ethics education, a further question arises: What form should this ethics education take?[7] I suggest below that based on the particular needs of science students and the learning outcomes that should be set for them, teaching activities should include role-play involving real cases or plausible hypothetical ones, with an emphasis on those situations with dual-use implications. In isolation these activities cannot meet all the needs of this cohort of students. However, as will become apparent through this chapter, they can make a significant contribution to this end, particularly if deployed in conjunction with other strategies, such as those aimed at improving student literacy.

3 In the *Critique of Pure Reason* Immanuel Kant famously drew attention to an important distinction between knowing a rule and knowing how to apply it. A134/B174. For instance a judge might have a good knowledge of the law but yet not know what law a particular case falls under, or a physician might be familiar with the descriptions of a disease but be unsure of the correct diagnosis when presented with a diseased patient.

4 For a good discussion of codes of conduct with respect to biological weapons, see Rappert, B. 2007, 'Codes of conduct and biological weapons: An in-process assessment' *Biosecurity and Bioterrorism: Biodefense Strategy, Practice, and Science*, vol. 5(2), pp. 145–54.

5 For a discussion of these factors, see Martin, B. 1992, 'Scientific fraud and the power structure of science', *Prometheus*, vol. 10(1), pp. 83–98; Hollander and Arenberg 2009, op. cit.

6 Good is being used here in the ethical sense, though it also has connotations related to validity, and there is frequently a connection between ethically and epistemologically good science, though not the space here to expand on this link.

7 The discussion here focuses especially on ethics education for undergraduate students, though some of the points made might be incorporated into training for practising scientists.

The Needs of Science Students

Many undergraduate science students experience particular difficulties when they undertake study in philosophical courses, even when these courses cater to their presumed interest in science (for instance, in classes on science and ethics or the philosophy of science). These difficulties are of concern not only because they limit student enjoyment of the subject studied, but also because evidence suggests students who have a negative orientation toward a subject will experience poor learning outcomes;[8] that is, they do not achieve what they ought to from their studies. Therefore, if the attainment of the generic and course-specific learning outcomes set for an ethics class is valued, these difficulties represent genuine concerns that need to be addressed.

Analysing the relatively limited literature regarding teaching philosophy to science students reveals that at the broadest level student difficulties stem from a difference in the culture and norms of the humanities and sciences. This manifests itself in science students frequently not having the requisite skills in writing, reading, and so on, to perform well in philosophy subjects generally (of which ethics forms a part); in their not knowing, understanding or being comfortable with the culture and expectations of philosophy; and in their possibly having a hostile orientation towards a discipline which they may perceive as either challenging or inferior to their chosen career path in science. These three specific issues demand special attention when teaching ethics to science students.

In expanding on the skills deficit experienced by science students, I want to begin by joining Geoffrey Cantor in observing that it is not simply the case that the difficulties encountered by many science students studying philosophy align with what he describes as the 'crude stereotype' that construes them as 'illiterate and culturally inept'.[9] Nor does the skills deficit result from a failure of intelligence or moral fibre; rather, it is frequently a product of the inexperience of students in certain types of activity, or occasionally a failure to value these activities. Sometimes it may also be the case that students have chosen a degree in science because they feel they lack natural competency in the skill set demanded by the humanities.[10]

The most widely discussed skills deficit canvassed in the higher-education teaching and learning literature in this area centres on essay writing. Essays are effectively an alien genre for many science students since they are generally

---

8 Murtonen, M. 2005, 'University students' research orientations: Do negative attitudes exist toward quantitative methods?', *Scandinavian Journal of Educational Research*, vol. 49(3), pp. 263–80.

9 Cantor, G. 2001, 'Teaching philosophy and HPS to science students', *PRS-LTSN Journal*, vol. 1(1), pp. 14–24, p. 15.

10 Cantor, for instance, makes this suggestion. Ibid., p. 20.

not required within the natural sciences and mathematics, though they may form part of the assessment in biology classes and in the so-called soft sciences such as psychology. According to Cantor, given that essay writing is not emphasised in science education, the 'prospect of essay writing may evoke fear and uncertainty since many science students will have no conception of what is involved or how to begin the process of essay writing'.[11] Problems divide into two main categories — those to do with an inability to engage in the analysis and thinking demanded by an essay, and those related to written expression. Into the first group fall the difficulties encountered by students with the very notion of developing and defending their own original point of view; with appreciating how to engage critically with the literature (including knowing how to analyse key concepts in written form); and with knowing what would constitute an appropriate answer to the question asked. The latter can manifest in students being unsure whether or not their essay even represents a legitimate response to the question.[12] Frequently such students submit papers that simply and unreflectively restate course content and border on plagiarism, since sources for the various arguments and concepts are not cited.[13] The second category involves the more general struggle of science students to write clearly and fluently. Such students also have problems knowing how to use the first person and active voice in essay writing, most likely as a result of being discouraged from doing so in scientific writing.[14]

If science students do not already possess or develop essay-writing skills they are unlikely to flourish in philosophy subjects which require them as part of their assessment. This means that not only will they achieve a poor grade in such subjects but they will fall down in their studies at a deeper level too. Sellars, for instance, makes the argument that essays are not merely arbitrarily associated with philosophy, 'they reflect the very nature of philosophy itself'.[15] If students cannot write an essay, they cannot 'do philosophy', they are not properly engaged in critical thinking, analysis and evaluation, and may be unable to construct an argument. On this view, philosophy teaching must incorporate the writing skills crucial to achieving critical analysis — the very stuff of philosophy. Lecturers and tutors should not regard their students as incompetent if they do not already possess writing skills, and the teaching and learning of writing should not be considered a merely remedial and distracting

---

11 Ibid.

12 Gooday, G. 'The Challenges of teaching history and philosophy of science, technology & medicine to "science" students', available: http://prs.heacademy.ac.uk/view.html/prsdocuments/66

13 Ibid. As Gooday notes, to many science students it would seem irrelevant or arrogant to attribute authorship to facts and theories in science.

14 Ibid.

15 This is Crome and Garfield's way of expressing Sellar's point in Crome, K. and Garfield, M. 2004, 'Text-based teaching and learning in philosophy', *Discourse: Learning and Teaching in Philosophical and Religious Studies*, vol. 3(2), pp. 114–30, p. 124.

activity the teacher is forced to engage in by virtue of student inadequacy.[16] As Sellars notes, it should be regarded 'an essential part of any training designed to teach students how to argue clearly and effectively. If our aim is to teach students how to think then we must accept that it will also be our task to teach them how to write'.[17]

Another feature of literacy increasingly acknowledged as deficient amongst the broader student population in universities is that of critical reading.[18] In the case of science students, problems may be compounded since many do not enjoy reading, are slow readers and they struggle to know how to evaluate a text.[19] In spite of this deficit, it appears little research has been undertaken into how to remedy this situation.[20] Yet, as with essay writing, an inability to read effectively is a significant impediment to studying philosophy, not just because it means students are unlikely to score well in their subjects, but also because critical reading is just part of what it is to do philosophy — to reflect on and analyse arguments and ideas. Thus, Kelton describes it as a 'gateway intellectual activity';[21] that is, an essential tool to getting started in one's philosophical studies. Crome and Garfield go further when they claim 'there is an intimate and unique bond between an appropriately engaged or active reading of a philosophical text and the act of doing philosophy itself'.[22] As with essay writing, these authors maintain that developing a student's reading skills should not be seen as a remedial activity, but part of the broader 'aim of all philosophy teaching, getting students to do philosophy'.[23]

There are two further skills that are important but often appear lacking amongst science students (again due mainly to lack of exposure and experience); namely, verbal ability and note-taking proficiency. Science students may not be comfortable expressing themselves in a public oral forum, and in lectures and tutorials they may struggle to know what is important and worth taking down, since they are accustomed to being provided with handouts or having key formulae clearly identified for them.[24]

---

16 There is a strange belief that teachers (including myself) sometimes have that students ought to already possess at least some of the skills and knowledge that we are charged with teaching them in a course. It is salutary to remind ourselves that our role as teachers is to teach, not complain when our students lack such knowledge.

17 Sellars, J. 2002, 'Some reflections on recent philosophy teaching scholarship', *PRS-LTSN Journal*, vol. 2(1), pp. 110–27, pp. 126–7.

18 Crome and Garfield 2004, op. cit. p. 115.

19 Ibid., pp. 119–20.

20 Ibid., p. 117.

21 Kelton, S. 1997, *On assessing philosophical literacy*, available: http://www.eric.ed.gov/ERICDocs/data/ericdocs2sql/content_storage_01/0000019b/80/16/c7/0d.pdf [viewed December 2009] p. 4.

22 Crome and Garfield 2004, op. cit. p. 122.

23 Ibid., p. 123.

24 Gooday, op. cit.

An inability to appreciate and embody in practice the different norms that govern the pedagogy of the humanities, as opposed to the sciences, contributes to the struggle with skills competency experienced by science students, particularly in essay writing. Gooday describes the problem well when he comments that '[w]hen learners enter into an unfamiliar field of knowledge, their entry is never *just* a simply undirectional process of picking up knowledge...novices need to secure the appropriate practices, strategies and expectations to be able to articulate and use such knowledge in accordance with the values of their specialist field'.[25] In the case of the sciences and humanities, the differences present a 'clash of cultures' in Gooday's view, which leaves science students unclear and confused over what is expected of them when studying philosophy. They may feel unsure of the rules of the game, do not necessarily understand what their teachers are seeking, and may find the tactics they have deployed to effect in their science subjects do not translate to success in philosophy. Again, as Gooday comments, '[f]rom the point of view of science students, the scholarly values of HPS [History and Philosophy of Science] teaching can seem bafflingly vague, gratuitously subjective and self-indulgent, whilst the pedagogical practices employed seem to lack a proper emphasis on "getting the right answer"'.[26]

This last point is significant and reflects a strong difference between the humanities and sciences cultures. From their mainstream courses, science students are reinforced in the belief (prevalent in broader society too) that there is just one truth of matter and it is the business of science to discover it. In this context the teacher and text are often regarded as authorities, and it is expected that there are definitive right and wrong answers to questions.[27] Therefore, to a science student the operation of an ethics class is highly puzzling. There is no one correct answer to ethical dilemmas so that neither textbooks nor lecturers are authoritative. Discussions are open-ended and can seemingly be mired in subjectivity and opinion. To succeed in the sciences students need to demonstrate they know and understand the dominant prevailing theory in a field, while in the humanities interpretative work is required and the ability to understand and critically evaluate a diversity of views including one's own. The ability to construct an original line of argument is also rewarded.

Related to the differences in culture and norms of the sciences and humanities (and again with the potential to hamper student learning) is the inadequate conception some science students hold of what the discipline of ethics is about.

25 Gooday, G. 2002, 'How do different student constituencies (not) learn the history and philosophy of their subject?' *PRS-LTSN Journal*, vol. 1(2), pp. 141–55, p. 147.

26 Gooday, op. cit.

27 Most philosophy of science since the influential work of Thomas Kuhn challenges the idea that science itself is actually like this. Nonetheless, these kinds of assumptions still appear prevalent in the teaching of science in higher education.

For instance, they may regard ethics as constituted by externally imposed rules and regulations; they may conflate ethics and law; or believe the discipline of research ethics exhausts the ethical issues raised by science, so that effective ethics committees may be all that is required to ensure ethical practice in science.

The final impediment to the learning of science students considered here is related to the hostility toward the entire discipline of philosophy (including ethics) found amongst many such students, their teachers[28] and scientific practitioners more broadly. Unfortunately, within this group, C. P. Snow's famous 'Two Cultures' thesis still appears widely accepted, namely that science and the humanities represent two quite separate cultures that lack even a common language to mediate between them. According to this division of the intellectual landscape, the sciences are superior to the humanities, with the latter often construed as irrelevant or just common sense. As former science 'insiders' who have gone on to work in philosophy, both Cantor and Gooday acknowledge the veracity of this perception. As Cantor describes it: 'When at school I shared with many of my peers the (utterly depressing) view that science students are innately superior to those taking humanities subjects, and that the sciences hold the key to the future.'[29] He goes on to discuss the self-selecting and mutually reinforcing nature of the community of science students who 'often perceive themselves as having chosen science and thereby positively rejected humanities subjects'.[30] They may regard the humanities in general as 'a doddle'[31] and philosophy, more specifically, as comprising waffle and navel gazing.[32]

If science students assume intellectual superiority over humanities students, any difficulties they experience with a philosophy class due, for instance, to the skills deficit described earlier will surely be particularly disturbing. They may well wonder how it is that armed with their natural academic ability they do not automatically prosper in their philosophical studies. Therefore, it seems probable that they will ascribe these difficulties to some fault with the teacher or the subject itself, further fuelling their frustration and antipathy toward the humanities.

Even if they do not have an overtly hostile orientation to philosophy, the study of applied ethics can be confronting for any student. As Joan Callahan comments, '[p]ractical ethics courses press students to become clear about their own biases

28 Part of the issue for this group may also be that they do not believe time in the curriculum should be devoted to a subject which is not strictly speaking science, or that such an ethical education is not needed because it is irrelevant or just common sense.

29 Cantor 2001, op. cit. pp. 15–6.

30 Ibid., p. 17.

31 Ibid., p. 18.

32 These observations are drawn from my own experience. I find such perceptions of philosophy particularly intriguing as they seem to me to be the antithesis of what philosophy is actually about; namely, rigorous argument well supported by evidence.

and to examine the reliability of their own ways of making moral decisions, and this, unavoidably, makes students feel vulnerable'.[33] Ethical positions are often deeply held, dependent on cultural and religious background, and may go unchallenged in daily life, so that formulating coherent reasons and justifications for beliefs may be difficult and intimidating for students. Many science students may also perceive philosophy of science and ethics as a direct threat to the discipline they have committed themselves to as students and potential future professionals. Unaccustomed by their scientific studies to engage in reflection on the philosophical basis for science and how it can be justified, or the ways in which it might legitimately be curtailed by the ethical concerns of society more broadly, they can interpret any such debate as a challenge to their personal integrity. The perception that the very existence of some philosophical disciplines represents an attack on the scientific enterprise and on individuals as participants in this enterprise is surprisingly widespread and continues into professional life. Frequently, scientists and those in medicine view discussion of ethics in these contexts as an unwarranted attack on their good intentions, calling into question their motives in a discipline they perceive to be dedicated to the public good. A philosopher's call to justify an ethical stance can be construed as unjust slander against one's person, partly because philosophers and ethicists are sometimes seen as outsiders who have no legitimate status or expertise with which to criticise the authority and status of science and scientists.

Whilst I think scientists, doctors and others may well be overreacting when they feel personally threatened and intimidated by philosophers, at the same time there is a sense in which education is and should be transformative — a catalyst to change. To reflect seriously and critically on one's world view may well be disconcerting and unsettling. A further unfortunate side-effect of such a transformation may be underperformance in science subjects, where the kind of scepticism and critical reflection encouraged in ethics and philosophy of science classes may undermine achievement in straight science subjects.[34]

Thus it seems from this critical examination of the literature that the challenge for teaching (and learning) ethics to science students is to investigate approaches which could build their skills, minimise experiences which may be threatening, while supporting the learning outcomes of philosophical ethics.

Although the difficulties experienced by science students are acknowledged in the literature and by teachers in the area, few systematic and well-researched solutions to such difficulties are proffered. Nigel Taylor makes this point when posing questions about possible principles to guide the teaching of philosophy

---

33 Callahan, J. 1998, 'From the "applied" to the practical: teaching ethics for use', in Kasachkoff, T. (ed.), *In the Socratic tradition essays on teaching philosophy*, Lanham: Rowman and Littlefield Publishers, Inc. pp. 57–69, p. 65.

34 Gooday, op. cit.

to non-philosophy students and the means of assessing good practice in this area: 'I shall not be drawing on any substantial body of literature about this topic, still less on an established body of "theory", for there isn't any.'[35] Therefore, he resorts to the same strategy as other authors confronting this problem; namely, he relies on critical reflection into his own experience, or self-reflection in tandem with the reflective experience of colleagues. The paucity of research in this area is not unique to the teaching and learning of non-philosophy students, but is part of a more general deficit in the philosophical literature. Sellars bemoans the proliferation of narrative, personal, experiential accounts of teaching philosophy, and hopes for more theoretical reflection in the future.[36] Part of the rationale behind this paper is to move beyond these kind of attempts, to systematically reflect on and re-conceptualise the problems, and address them by appealing to the broader literature in higher-education teaching and learning.

Setting the Ethical Agenda

Having established that science students need ethics education (bearing in mind they are also a group with special needs when studying humanities), what goals should be set for their ethical education? In other words, what learning outcomes — skills, knowledge, and so on, should students take away from ethics classes?

The 'virtue/skill dichotomy' represents a fundamental divide in approaches to ethics education of medical students that is also relevant in this context.[37] Those who emphasise the virtue approach claim that ethical instruction should teach virtue, students should exit classes as better people, and courses should transform them into citizens of good character. In a tradition originating with the ancient Greeks, the underpinning rationale is simple — that good people make good decisions.[38] Conversely, those who lobby for a skills focus argue that the emphasis in ethics education should be on fostering skills and resources required to engage in solving ethical problems. Whilst the virtue position embodies a noble and laudable goal that may follow from ethics education, I agree with Eckles *et al.* that it must surely lie beyond the primary business

35 Taylor, N. 2003, 'Teaching philosophy to non-philosophy students: The example of architecture and town planning', *Discourse: Learning and Teaching in Philosophical and Religious Studies*, vol. 3(1), pp. 41–52, p. 41.

36 Sellars 2002, op. cit.

37 The literature on teaching medical students ethics is more developed than that for science students.

38 Justin Oakley has argued for the virtue approach to be adopted in teaching science students, citing the experience with medical students to support his case. Oakley, J. 2009, 'Teaching scientists the value of virtue', *Australasian Science*, vol. 30(2), p. 39.

of ethics teachers.[39] Such a learning outcome would also present significant challenges for assessment. How could a teacher determine whether or not students had attained this learning outcome? What possible measure could be used to decide if they were of good character? Therefore, given the problems inherent in the virtue approach it seems more plausible to accept the slightly less lofty, skill-based goal for ethics education of science students. Thus, at the broadest level, the aim should be to teach students the skills that will equip them to practise science ethically. To achieve this aspiration I would argue that there are two groups of learning outcomes that should be set — those to do with understanding the nature and terrain of ethics, and others related to how to do ethics or applied philosophy more generally.

Students need to develop an appreciation of some of the key features of ethics in order to operate ethically in science. In the first instance, they need to acquire what I call their 'ethical radar'; that is, a sensitivity that allows them to recognise moral issues, as well as situations where values are in conflict. Such an understanding may appear common sense and such conflicts might seem obvious, but this, in my experience, is not necessarily the case. For instance, some students in medical sciences consistently conflate ethical and clinical choices, and assume they are making purely medical decisions based on scientific factors when their conclusions are being driven by their own personal values (for example, what constitutes a good quality of life). Similarly, it may be difficult for a student in the life sciences to see that there is any ethical issue generated by research into a deadly virus because they may implicitly value scientific knowledge, research and progress over issues such as threats to public health and security. Thus, they might be unconcerned by research into transmission of the Ebola virus, for instance, seeing the science as neutral or focusing on the good ends such research might produce, rather than its potential to lead to a disease outbreak or biological attack.

Related to developing sensitivity to the presence of ethical issues is the goal of learning to appreciate the complex and contested nature of the ethical landscape. Students need to understand that there are genuine and well-grounded points of difference between people on such issues, but that although complete consensus on the nature and resolution of a concern may be rare, this does not undermine the value of discussion and argument, nor does it mean that ethical relativism must follow.

---

39 Eckles, R. E., Meslin, E. M., Gaffney, M. and Helft, P. R. 2005, 'Medical ethics education: Where are we? Where should we be going? A Review', *Academic Medicine*, vol. 80(12), pp. 1143–52.

To complement this appreciation of the nature of ethics, students should acquire a particular set of skills regarding criticism, analysis and argument. Although discussing philosophical pedagogy more broadly, the aims Sellars sets down apply equally well in practical ethics. He writes:

> If one defines philosophy as a critical analysis of one's existing opinions and the attempt to replace those opinions with rationally ground beliefs, then teaching philosophy should involve teaching the skills necessary to accomplish this. A successful philosophical education, then, will be one at the end of which one's students are able to call into question their own unexamined presuppositions and to think rationally for themselves.[40]

Students need to be able to interrogate and critique ethical views expressed in the literature, in the media and by their colleagues, as well as systematically reflect on their own views. Students should be able to argue and formulate coherent, reflective and well-justified responses to ethical situations, both verbally and in written form. Therefore, it is insufficient to accept the UK Quality Assurance Agency for Higher Education's claim that in the biosciences we should expect all students to develop 'some understanding of ethical issues and the impact on society of advances in the biosciences', but that only good students should 'be able to construct reasoned arguments to support their position on the ethical and social impact of advances in the biosciences'.[41] Good students will presumably produce better arguments than poorer students, but to have no expectation that poorer students will develop any skills in argument is to set the bar too low. If students do not obtain competency in such a fundamental learning outcome, they should not pass the course in which they are enrolled.

Further learning outcomes have been suggested in the literature, including that ethics education should canvass the 'development of competencies in analysing how social and technical factors interact'.[42] I would argue, however, that this is beyond the purview of philosophical ethics and that traditional teachers of ethics would likely lack the relevant expertise to do justice to such a goal, though if such outcomes are regarded desirable by some institutions, then teachers should be sourced and course goals amended.

---

40 Sellars 2002, op. cit., p. 126.

41 As reported in Willmott, C. J. R., Bond, A. N., Bryant, J. A., Maw, S. J., Sears, H. J. and Wilson, J. M. 2004, 'Teaching ethics to bioscience students — A Survey of undergraduate provision', *BEE*-j, vol. 3.

42 Hollander and Arenberg 2009, op. cit. p. 11.

Meeting the Challenges

Pedagogy is important. There are better and worse ways to teach particular subjects and disciplines. One factor demonstrated to benefit student learning in all areas of higher education is the constructive alignment of learning activities (that is, the activities in which students engage during their studies) with outcomes and assessment.[43] In essence, constructive alignment ensures that what is done in the classroom supports what students ought to learn. It involves determining appropriate learning outcomes, having students perform tasks and activities that develop these outcomes (often particular skills or abilities), and rewarding successful achievement of goals by awarding marks. For instance, if the goal is to teach medical professionals how to give injections, rather than delivering lectures on how to do this and assessing their skills by a written paper, students should practise this skill (initially in some form of simulated environment) and teachers should assess by valuing it relative to other learning outcomes by assigning a particular grade. Such a strategy might sound self-evident; however, for various reasons including historical and pragmatic ones, such an obvious strategy may not always be adopted. For instance, Kelton discusses how he used multiple-choice exams with computer grading, partly as a way of coping with the enormous volume of marking generated by large class sizes. However, eventually he came to regard such tests as ineffective in assessing whether students had achieved learning outcomes.[44] Courses and entire disciplines might also have certain assessment strategies historically linked to them that may not align with or be the best way of ensuring the attainment of learning outcomes.

Now I want to put the case for why role-play (focusing especially on dual-use dilemmas) can help meet some of the important learning outcomes identified for science students in ethics. First a couple of the key terms need to be defined. By 'role-play' I mean a structured exercise in which participants are assigned roles and some form of scenario in which to play out these roles.[45] For the purposes here, a dual-use dilemma is a situation that arises when one and the same technology, scientific research project or outcome of a scientific research project is such that it can be used as a basis to provide means to significantly

43 Biggs, J. 1996, 'Enhancing teaching through constructive alignment', *Higher Education*, vol. 32, pp. 1–18; Biggs, J. 1999, *Teaching for Quality Learning at University*, Buckingham, UK: SRHE & Open University Press; Prosser, M. and Trigwell, K. 1999, Understanding learning and teaching: the experience in higher education, Buckingham, UK: SRHE & Open University Press; Ramsden, P. 2003, *Learning to Teach in Higher Education*, 2nd edition, London: Routledge.

44 Kelton 1997, op. cit.

45 For a good example of a dual-use role-play, see Rappert, B. 'The Life sciences, biosecurity and dual use research', available: http://proejcts.exeter.ac.uk/codesofconduct/BiosecruitySeminar/Education/index.htm [viewed December 2008].

harm others as well as perform another purpose that is not harmful. A dilemma arises in such a case, as there are reasons both for and against developing the technology or conducting the research.[46]

Evidence suggests that role-play with dual-use cases is a highly effective teaching activity to support the learning outcomes outlined earlier. In the first instance, role-play can help students develop their ethical radar, enabling them to become aware of the issues generated in scenarios.[47] Such scenarios can also foster an appreciation of the complex and contested nature of practical ethics generally. In applied ethics, opinions can differ over what values are at stake, and what the morally correct response should be. Since role-play can force students to adopt and defend positions they may not actually hold, and engage with other similarly positioned individuals, they learn to appreciate that there can be multiple legitimate positions to any ethical debate.[48] Students have reported that role-play may be superior to other teaching and learning strategies when it comes to developing a sense of the ethical terrain. They noted that in role-play they could '[c]reate a discussion...[which] makes you transport yourself to the role and situation...[and] [s]ee it from "different shoes"' and 'It makes people think, adopt different points of view, and therefore, get a broader understanding of an issue.'[49]

To assist psychology students in learning about the complexity of research ethics and demonstrate that there are multiple perspectives that should be considered in the evaluation of studies, Rosnow developed a role-play exercise.[50] Strohmetz and Skleder later evaluated the effectiveness of this role-play in achieving the stated learning outcomes, validating Rosnow's work.[51] Other authors have also supported the value of role-play. Chesler and Fox, for instance, 'suggest that students can achieve insights into themselves, others, and motivations for actions which "can aid students in clarifying their own values and in effectively

46 Definition adapted from unpublished research by Dr. John Forge, University of Sydney.

47 For instance, Illingworth argues role-play can 'enrich students' perceptions of morally significant situations', Illingworth, S. 2004, *Approaches to ethics in higher education: Teaching ethics across the curriculum* University of Leeds: Philosophical and Religious Studies Subject Centre, p. 52; Brummel, B. J., Gunsalus, C. K., Kristich, K. L. and Loui, M. C. 2008, 'Development of role-play scenarios for teaching responsible conduct of research', available: http://netfiles.uiuc.edu/loui/wwn/RCRRolePlays.pdf [viewed December 2009], p. 1.

48 Brummel *et al.*, Ibid.

49 Ibid., p. 6.

50 Rosnov, R. L. 1990, 'Teaching research ethics through role-play and discussion', *Teaching of Psychology*, vol. 17(3), pp. 179–81, p. 179.

51 Strohmetz, D. B. and Skleder, A. A. 1992, 'The Use of role-play in teaching research ethics: A validation study' *Teaching of Psychology*, vol. 19(2), pp. 106–8, p. 108.

directing or changing their own behaviour'''.[52] Similarly, Doron comments that role-play 'exposes the students to attitudes or viewpoints that they might otherwise not have been conscious of'.[53]

In addition to helping develop their ethical radar, role-play can assist students to pick up analytic, critical and argumentative skills. Primarily through their deployment in simulated settings, students learn the skills required to navigate real ethical situations. The task they embark on in role-play is not abstract and theoretical but highly immediate and practical. By being forced to understand, interrogate and attack one's opponents and defend a position in the cut and thrust of debate, a whole raft of skills may be developed.

Significantly, the use of role-play can also be an effective means of dealing with some of the special needs of science students in the ethics classroom. Engagement in role-play can break down the hostility toward ethics sometimes encountered, by providing an enjoyable and stimulating environment for learning. Brummel, for instance, reports on the high level of student satisfaction associated with this form of learning activity,[54] and Doron comments that role-play 'facilitates deeper individual involvement with, and interest in, the case' being examined.[55] As the higher-education literature has shown, this is significant, because positive learning experiences can translate into better learning outcomes.

The potentially threatening nature of ethical discourse for science students can also be ameliorated by role-play since, as Brown notes, such scenarios 'have the advantages of creating low-risk conditions for expression of extreme opinions by students'.[56] He goes on: 'The freedom afforded by playing a stranger, and attributing extreme positions to that individual, allows the players tremendous scope of exploration into the nuances and conflicts inherent in any complex situation, without exposing the players' own beliefs.'[57] However, Brummel notes there can also be unfavourable responses to role-play, with students potentially feeling awkward or not seriously engaging with the exercise.[58] In his research, such shortcomings did not outweigh perceived advantages on the part of students generally and, in my view, could potentially be handled by sensitivity on the part of the teacher to the needs of students who are not as socially competent or skilled as their colleagues.[59]

---

52 Quoted in Brown, K. M. 1994, 'Using role play to integrate ethics into the business curriculum a financial management example', *Journal of Business Ethics*, vol. 13(2), pp. 105–10, p. 105.

53 Doron, I. 2007, 'Court of ethics: Teaching ethics and ageing by means of role-playing', *Educational Gerontology*, vol. 33(9), pp. 737–58, p. 742.

54 Brummel *et al*. 2008, op. cit., pp. 2 and 4.

55 Doron 2007, op. cit., p. 742.

56 Brown 1994, op. cit., p. 105.

57 Ibid., p. 106.

58 Brummel *et al*. 2008, op. cit., p.6.

59 Illingworth gives some excellent and practical suggestions for teaching ethics generally which are particularly relevant in this context. She focuses specifically on creating a safe environment, ensuring mutual

Role-play also enables students to be immersed in an environment where the potentially alien norms of the humanities operate and can be implicitly fostered. A very different notion of the teacher prevails in such scenarios. For instance, the manner in which role-plays function mean they 'avoid...preaching by the authority figure'.[60] Doron supports such a view, claiming that in role-play the teacher 'is not the omniscient expert who possess the correct answer and whose place in the classroom is laid down by a hierarchy of superiority that is based on the disparity between the teachers' and their students' knowledge and experience'.[61] According to Brown, another advantage of such exercises is that '[b]y its nature, a role-play has no ultimate solution contained within it, and so emphasises the indeterminate elements of decisions'.[62] There is no single correct answer that is possessed by authorities, either teachers or texts.

Dual-use cases add to the effectiveness of role-play by helping provide the realistic and practical context science students crave. As recent research with ethics teachers in science has shown, '[w]hen it comes to the aspects of ethics respondents believe students find most interesting or engaging, the most prominent theme is that of real world cases, or ethics in social contexts'.[63] Not only do real dual-use scenarios provide a realistic setting for ethical discussion, they also have at their heart a significant conflict, which as Brown has argued, is essential for the success of role-plays.[64] Dual-use situations have an inherent tension between values — the desire for scientific knowledge and progress, and the potential for such knowledge to generate harm. Finally, dual-use scenarios demonstrate vividly to science students that even if they are well-intentioned scientists and moral people more generally, their research can still create significant ethical concerns. In the heat of role-play they learn that they need to deal with these ethical concerns if they are to be effective scientists in the twenty-first century.

Conclusion

Though it may represent a challenging exercise, science students need to be taught ethics. They need to be empowered with the skills to conduct their professional lives in the face of the moral challenges they will confront. Role-plays based on dual-use dilemmas can motivate engagement with ethics, be a

respect and protecting confidentiality. Illingworth 2004, op. cit., pp. 82–90.

60 Brown 1994, op. cit., p. 105.

61 Doron 2007, op. cit., p. 743.

62 Brown 1994, op. cit., p. 106.

63 van Leeuwen, B., Lamberts, R., Newitt, P. and Errington, S. 2007, 'Ethics, issues and consequences: conceptual challenges in science education', in UniServe Science Teaching and Learning Research Proceedings, available: http://science.uniserve.edu.au/pubs/procs/2007/23.pdf [viewed December 2009] pp. 112–9, p. 117.

64 Brown 1994, op. cit., p. 106.

catalyst to developing critical, analytic, argumentative and verbal skills, and do so in an enjoyable and non-threatening way, conducive to getting the best from current students and future scientists.

Conclusion: Lessons for Moving Ahead

BRIAN RAPPERT AND LOUISE BEZUIDENHOUT

If we are to avoid the life sciences becoming the death sciences — as has happened in so many fields of knowledge — then concerted thought and action is required. *Education and Ethics in the Life Sciences* has attended to one aspect of the 'web'[1] of measures necessary to avert this prospect. The need for greater education and awareness about the security–science link was one of the reasons motivating this volume. Its contributions have supported this starting impetus and elaborated the contours. The authors have examined a variety of emerging efforts to attend to the possibility that the life sciences might aid in the spread of disease; most of which addressed issues far beyond sturdy locks on laboratory doors. An aim has been to share experiences, models and resources with readers.

Experiences: Selgelid, Sture, Johnson, and Barr and Zhang have asked how ethics — and particularly ethics training — has been and could be brought to bear on dealing with so-called dual-use concerns. Each of these contributors, as well as others, have cautioned against thinking that the destructive use of science could be addressed by simply requiring budding researchers-to-be to sit in general ethics modules during their university degrees. Whether because of the hidden curricula that often frustrate teaching leading to principled outcomes, the skewed past priorities of bioethics, the need for ongoing and workplace-relevant instruction, or the resistance given to formal ethical instruction, many of the authors have detailed the vital importance of measured and context-sensitive interventions. As argued by Barr and Zhang in particular, despite the international character of much of the life sciences, the national structures in place for research and bioethics are of major significance. Not only do research cultures differ, but the pressures on the individual scientist can also vary dramatically between countries and research communities. It is therefore of marked importance that dual-use education not only address the ethical background of scientists, but also the social and cultural context in which they are operating. In developing countries, for example, considerations such

1 See Rappert, B. and McLeish, C. (eds) 2007, *A Web of prevention: Biological weapons, life sciences and the governance of research*, London: Earthscan.

as limited resources, governmental and social expectations, as well as cultural norms may affect the way in which ethical education is internalised and applied (see below).

Models: The chapters — particularly those in Part 2 — outlined a range of possible rationales, strategies and methods for bringing in educational measures. The inter-related chapters of Minehata and Shinomiya and Mancini and Revill showed how a process of 'survey-contact-network-assist-resurvey' provided a basis for building national understanding and interest. Enemark described a series of regional workshops in Australia conducted by members of the National Centre for Biosecurity. Drawing on past experiences elsewhere, the Centre members were able to complement renewed concern about the control of sensitive agents and devise an agenda for required future activities. Because of their long-term engagement with biosecurity-related issues, Connell and McCluskey suggested a number of routes for introducing attention to the misuse of science into a university setting: 'Responsible Conduct of Research' training, institutional biosafety committees, laboratory-safety training, a biodefence certificate, and an institutionally based 'train-the-trainer' system. As they argue at the end of their chapter, the strengths, limitations and prospects for each of these approaches needs to be seen against the consuming time demands placed on those associated with the life sciences. At the level of national governments, Garraux gave a work-in-progress account of the implementation of an awareness-raising project in Switzerland; one that suggests how those inside and outside of government can work together in mutually beneficial ways. His lessons outline paths for sustaining long-term outreach to both academic circles and first responders. Finally, Friedman detailed the most 'top-down' example of an education development given in this book. The Israeli case illustrates how science and security organisations can work together to establish a national structure for advancing education. Despite focusing on particular initiatives associated with their own work, each of the authors suggested that a multi-level approach is needed to address biosecurity education.

Resources: The Education Module Resource (EMR) described by Dando and Whitby presents easily accessible, electronic-support material that can be used to raise awareness. By avoiding the 'one-size-fits-all' approach, this resource is available for lecturers to fit into existing programmes (as further discussed by Revill and Mancini). Dando and Whitby also forwarded an expert-level distance-learning programme that makes use of innovative electronic online technologies that facilitate outreach on a worldwide basis. In her essay on teaching ethics to science students, Johnson not only highlighted the importance of pedagogy and dedicated teaching of ethics for science students, but also gave an outline of innovative methods such as role-playing which make ethics more accessible to students.[2]

So in summary, *Education and Ethics in the Life Sciences* has indicated possible guiding philosophies, strategies, enabling mechanisms, techniques and materials. But just as the contributors have detailed the importance of context-sensitive tuition, it needs to be recognised that the achievements documented in this volume cannot be reproduced elsewhere by a simple mechanical duplication. The arguments offered are given in the spirit of providing a springboard for creative thinking and action. Their relevance should be interpreted in different ways across varied situations, a point taken up shortly.

The need for active questioning of the significance of lessons from this volume (and elsewhere) is underscored by a point made by Dando and Whitby: while it is possible to demonstrate how some forms of education and ethics training help further specific teaching objectives, there is little in the way of agreed standards for evaluating the significance of ethics teaching on future behaviour.[3] Moreover, as Connell and McCluskey note, studies that have been done of major ethics-related training requirements — such as those associated with the US National Institute of Health's (NIH) Responsible Conduct of Research initiative — indicate only modest accomplishments in relation to the criteria used to measure them. Therefore, caution is prudent in thinking about what works and what can work.

Arguably, these points also allude to a fundamental question that has reappeared throughout this book; that being: 'What must be done and by whom?' As included in the Introduction, what counts as appropriate education in areas such as biosecurity is often contested because what is considered suitable by the way of education in general is often debated. In her chapter, for example, Johnson wrote of the basic divide between virtues and skill ethics, as well as the complications of philosophers teaching science students. More generally, notions about proper education are inextricably bound with the exercise of authority and expertise. As such, various types of education are not just different ways of achieving the same goal, but themselves are tied to alternative answers to the aforementioned question.

The scope for contention is all the greater in relation to matters central to *Education and Ethics in the Life Sciences*. As with other aspects of security, just

2 As considered as well in Rappert, B., Chevrier, M. and Dando, M. 2006, *In-depth implementation of the BTWC: Education and outreach*, Bradford Review Conference Paper No. 18, available: http://www.brad.ac.uk/acad/sbtwc/ [viewed 1 April 2010]; and see: http://projects.exeter.ac.uk/codesofconduct/BiosecuritySeminar/Education/index.htm.

3 Derived from a reading of National Academies 2009, *Ethics education and scientific and engineering research: What's been learned? What should be done?* Washington, DC: National Academies Press.

what 'biosecurity' should mean and how it can be achieved are matters on which individuals differ. Security, for whom and from what, are only some of the topics leading to division. Today, much of the government-level interest in biological weapons relates to sub-state terrorist groups. While some of those in this volume expressed concern in this regard, others did not. Indeed, many of the senior contributors have been working on bioweapon issues long before the recent upturn in attention to them. Much of the scepticism about the terrorist threat derives from doubts about the ease of inflicting mass casualties through disease. Attention is directed instead towards maintaining the resilience of the existing stigma and prohibition against the deliberate spread of disease into the future — whatever that future brings by the way of new technological possibilities and security environments.

Here, as elsewhere in matters of public policy, the 'process of formulating the problem and of conceiving a solution (or re-solution) are identical, since every specification of the problem is a specification of the direction in which a treatment is considered'.[4] In this regard, it is worth noting that discussions to date about what needs to be done to prevent the destructive application of the life sciences have overwhelmingly been couched within traditional national-security frameworks. Such contexts have stressed the importance of limiting access to materials, agents, findings, equipment, and techniques.[5] However, it could be argued instead that the question at hand should be one of how modern overall science can be entrusted to improve societal wellbeing and security — rather than control them, the focus would be with making science relevant to societal needs. Tackling this would require addressing how social trust is engendered in science and its institutions. That, in turn, is not a matter for technical or policy experts only. Rather, it requires a much more inclusive societal discussion.

While contributors in this volume have written about the need for ethicists, social scientists, and government officials to re-think their commitments and practices, the majority of attention has been directed towards those associated with the life sciences. In turn, within this diverse group, 'scientists' have been at the centre of the discussion about education. That has been justified largely on the basis of the current lack of professional attention to the hostile use of the life sciences, the need for practitioners to become involved if sensible responses are to be devised, and the potential contribution to maintaining the current prohibition. However, clearly the concerns that motivated *Education and Ethics in the Life Sciences* are not matters for scientists alone.

4 Rittel, H. and Webber, M. 1973, 'Dilemmas in a general theory of planning', *Policy Sciences*, Vol. 4, pp. 155–69.

5 For further elaboration of the limits of current framing of biosecurity, see McLeish, C. 2007, 'Reflecting on the Problem of Dual Use', in Rappert and McLeish (eds), op. cit.; Vogel, K. 2008, 'Framing biosecurity', *Science and Public Policy*, vol. 35(1), pp. 45–54.

Therefore, it follows from the previous three paragraphs that standards for assessing the significance of teaching can be an issue on which people disagree. Even if relative accord were reached on what has been taken as a central aim in this book — namely, preventing the hostile use of the life sciences — measuring the contribution of any education effort to this goal would be problematic. Although it is possible to use certain metrics for evaluating education (such as improved knowledge and understanding of individuals, and the satisfaction of participants with training), these are very much secondary, proxy measures in relation to this objective. Such points along with others from this chapter suggest the need for vigilance regarding what counts as effective learning.[6]

So, while proposing models and resources for further efforts to extend education, in doing so the contributions to this volume have underscored many areas for work that remain. In closing, it is possible to suggest three of these.

1. Widening Engagement: In their study on bioethics education in China, Barr and Zhang mentioned what they called a 'software' problem, meaning that while the facility design met with the required guidelines, the human element of biosecurity had been neglected. When considering extending engagement with issues of education, it is important to recognise the scenarios in which more attention has been spent on developing the facility security than training the staff within it. This is a particularly pertinent situation for resource-limited countries needing to maximise funding results. Including such countries in science-security debates requires sensitivity towards the potential pressures of limited physical and human resources as well as different social priorities.[7]

While US and European academia heavily influence life-science research around the world, it is important that each country examines its own ethical tradition and how it is unique. Education cannot be merely a case of pushing foreign solutions onto life scientists, but should be sensitive to their pressures and beliefs. Scientists, science students and others related to the life sciences do not present a homogenous group of ethical and cultural affiliations, a fact which should be reflected in online repositories. While lack of previous engagement is an international challenge, it has the potential to be rendered more acute by a lack of culturally appropriate teaching methods and content. Further discussion on innovative teaching methods and contextually appropriate case studies will no doubt strengthen and support pedagogy in developing countries.

In need of further acknowledgement is the recognition that differing access to resources such as the internet (for online learning and discussion groups) may

---

6 For a consideration of this theme, see Argyris, C. 2003, 'A life full of learning', *Organizational Studies*, vol. 24(7), pp. 1178–92.

7 On the latter, see Gould, C. 2009, 'Conclusion', in Rappert, B. and Gould, C. (eds), *Biosecurity*, London: Palgrave.

affect the ease with which initiatives are accessed and utilised. It is important that international education projects consider the possible limitations of their choice of technology for certain nations. As developing countries are more likely to rely on the efforts of champions than their developed counterparts, the significance of support and training for invested teachers cannot be overemphasised.

Dando and Whitby also suggest the development of national and regional networks that may allow much faster development and uptake of material suitable for different countries and regions. This has the potential to become a very powerful resource, as local networks would allow issues to be discussed against a more similar ethical background. It would also facilitate the development and sharing of educational resources that may reflect topical issues and context-appropriate examples. Networking and collaborative partnerships have been strongly encouraged, for example, to develop science and technological innovation in sub-Saharan Africa, and it is possible that the support and exploitation of existing and future research networks could be a valuable tool for expanding the debate.

The process by which policy is developed in individual countries differs, as do the range of people involved. Furthermore, the level of public participation in policy development varies greatly between nations. Nonetheless, there is an international trend towards developing plans on biosecurity in which scientific communities could play a major role. As mentioned by Christian Enemark in his chapter, 'by becoming more familiar with political and policymaking processes, life scientists might be able to suggest better ways of managing the security risks inherent in some research while minimising scientific opportunity costs'. This is of vital importance to countries with emerging science-research communities. Additionally, with a growing number of collaborations between developing and developed countries it is becoming increasingly important that scientists from the former have a voice not only in their national debates, but also in international research communities. This is necessary in order to have their particular situations recognised and considered.

2. Ethics as a Non-issue: As suggested in the Introduction, one the curiosities of the contemporary discussion about the security dimensions of the life sciences is that while it is often said that any knowledge might be used for destructive ends, in practice it has been extremely rare that benignly intended civilian research has been identified as posing concerns.[8] Arguably this speaks to the conceptually confused manner in which the said 'dual-use' potential of

8 Rappert, B. 2008, 'The risks, benefits, and threats of biotechnology', *Science and Public Policy,* February, vol. 35(1), pp. 37–44.

knowledge is conceived. The chapters in *Education and Ethics in the Life Sciences* have indicated the lack of previous engagement with security-related issues among practitioners across varied national contexts.

This overall situation raises practical and conceptual questions regarding how the security potential of science are and could be assessed. Moving into the future, it would seem highly prudent to better understand how, for whom, between whom, and under what circumstances the implications of research become matters of concern. One way forward along these lines would be to examine why practitioners have not identified the potential for destructive application of research.[9] In other words, what institutional structures, professional preoccupations, or other factors, have rendered concerns about the damaging use of the life science 'non-issues' for so many? Such an approach would presumably inform discussions about what kind of education would be relevant to practitioners.

3. *International Leadership:* Practically, it is clear that nascent efforts could be strengthened through international co-ordination and leadership. What is needed is a forum for building high-level agreement about what should be done and sharing experiences.

As a key cornerstone of the prohibition and stigmatisation of biological weapons, the Biological and Toxin Weapons Convention (BTWC) could help fulfil such roles in the future. As noted in this volume, education has been identified as an important topic between governments party to the treaty in recent years. Indeed, as Dando and Whitby document, explicit international recognition of the significance of the awareness and education of life scientists runs back to the Second BTWC Review Conference in 1986. However, as they also suggest, this recognition in itself has not been sufficient to deliver adequate concrete measures. Therefore, what is needed is a plan for concerted action. That plan could include mutual targets, deadlines, and milestones; the establishment of international and/or regional co-ordinators; a programme of international workshops; and agreed bilateral and multilateral assistance.

Perhaps what is needed most for the future is a collective vision among states, intergovernmental organisations and members of civilian society. As part of an address to the Meeting of Experts of the BTWC in 2008, one of the authors (Rappert) proposed such elements for a shared vision, including agreeing:

> ...all those graduating from higher education in fields associated with the life sciences should be familiar with the international prohibition against biological weapons;

9 Eliasoph, N. 1998, *Avoiding politics*, Cambridge: Cambridge University Press.

> ...all those undertaking professional research careers should have received effective training or instruction related to preventing the misuse of their research;
>
> ...each government should commit itself to initiating a dialogue with their respective national science academies (or other relevant bodies) about how the present low level of awareness can be swiftly corrected.

Regardless of whether or not these elements are proper, current discussions and efforts would do well to organise themselves around such positive shared goals that are subject to joint questioning.

As outlined in the Introduction, if handled properly, concerted attention and action to education in the BTWC could help ensure that conventions remain meaningful and robust moving ahead. The 2011 BTWC Review Conference provides an opportunity to formulate and arrange such a plan; one that should not be lost. The formation of such a shared agenda between governments could open a new chapter of reflection on how to ensure the prohibition of biological weapons.

www.ingramcontent.com/pod-product-compliance
Lightning Source LLC
Chambersburg PA
CBHW060929170426
43192CB00031B/2877
* 9 7 8 1 9 2 1 6 6 6 3 8 4 *